BRAD STEIGER

A ROADMAP OF TIME

How the Maxwell/Wheeler Weather-Energy Cycles Predict the "History" of the Next 25 Years

PRENTICE-HALL, INC.
Englewood Cliffs, New Jersey

A Roadmap of Time

Printed in the United States of America
Prentice-Hall International, Inc., London
Prentice-Hall of Australia, Pty. Ltd., Sydney
Prentice-Hall of Canada, Ltd., Toronto
Prentice-Hall of India Private Ltd., New Delhi
Prentice-Hall of Japan, Inc., Tokyo

10 9 8 7 6 5 4 3 2 1

Library of Congress Cataloging in Publication Data
Steiger, Brad.
A roadmap of time.
Includes index.
1. Meteorology—Periodicity. 2. Weather forecasting. 3. Climatology. I. Title.
QC883.2.C5S74 551.6'365 75-22259
ISBN 0-13-781385-6

Acknowledgments

The author wishes to thank John Cejka, president of Cyclomatic Engineering, Inc., for his invaluable assistance in the compilation of the research materials necessary to write this book. I am also grateful to John for serving as my own guide to a deeper and more complete understanding of the Roadmap of Time.

Inquiries concerning the work of Cyclomatic Engineering should be directed to Post Office Box #382, Glenview, Illinois 60025.

Contents

PART III *THE NEXT 25 YEARS–AND HOW TO LIVE THROUGH THEM*

PART

DISCOVERY

1

Cyclomatic Engineering, Inc.

Such things as stock and commodity prices, industrial production, war, depressions, civil riots, and stock-market crashes are considered blights which occur without structure or demonstrable pattern. But man's science is founded on the repeatability of experiments. His religion is built on the promise of plan and order. Something within man seems to drive him to find patterns in his world, his universe.

In the 1500s the famous French jurist Jean Bodin found that the northern (cool climate) races were more faithful to their governments and less fanatic in their religion. They were more tolerant and docile, happier, and more trustworthy. On the other hand, the more southern and warmer races were more melancholic in disposition, more malicious, less democratic, more given over to the institution of slavery, and more dependent upon tyrants and dictators.

Back around 320 B.C. the Greek philosopher and scientist Aristotle, tutor of Alexander the Great, made many of these same observations.

Now in the 1970s we have permitted ourselves the luxury of believing that we have largely emancipated ourselves from the forces of environment. We live in temperature-controlled homes, in which climate has been stabilized. Gravity bothers us very little, for our modern inventions permit us to circumvent its effects with ease. Although once in a great while someone may get struck by lightning, consider the many ways in which we have succeeded in harnessing the force of electricity.

But an increasing amount of evidence has been steadily accumulating that man is nowhere as emancipated from his environment as he has thought.

From many recognized studies, we can gather the following information: When temperatures are excessively high and the barometer is going down, people become depressed, dull, tired, and absentminded. More articles are left in stores or on trains and buses in hot, humid weather. People are more forgetful, more irritable, less tolerant of one another. Their nerves are on edge. People are more pessimistic; they have less self-confidence and "drive." They are more impulsive and subject to compulsive behavior.

At such times more accidents occur on the street, in cars, in homes, in shops, and in factories. When it is hot and humid, you are more likely to cut your thumb or break a dish. A factory worker is more likely to get caught in machinery. Coordination is poorer. Typists will hit more wrong keys and make more mistakes in their spelling.

During hot and humid weather there are more violent episodes among the patients in mental hospitals. Mentally normal patients in general hospitals do not feel as well. More well people become sick, more sick people die.

Children are restless in school; they cannot concentrate; their minds wander; they are more irritable and less obedient and cooperative. Teachers are lethargic, less patient and understanding.

More people faint in hot, muggy weather. People cannot hold

their liquor as well. (Warm-climate people drink less than cold-climate people for the same reason.)

If you can help it, do not go to the dentist on a hot, muggy afternoon. The drill will hurt more. Don't make an important sales call then, for you will meet with greater resistance. People are less suggestible, more prone to find fault, and more likely to be suspicious.

Decisions of executives are not likely to be as good, for people are then more erratic and less careful in their thinking. They are more emotional and less rational. When civil service tests are taken in August, only 58 percent pass. Seventy-five percent of the applicants are successful when they take the test in April or November.

Sixty-nine percent more New Yorkers are arrested for assault and battery when the temperature is between 80° and 85° than on cool days. Crimes against persons are of greater frequency and violence during warm weather, warm seasons, years, or decades. There is more crime against property during cool times.

But how far can these observations take us?

Astronomer Selby Maxwell, once science editor of the Chicago *Tribune*, discovered a weather-energy cycle which has proved to be the basic cycle which governs *all* weather—past, present, and future. Maxwell determined the correct time lags which cause the turbulent upper air masses to act in a predetermined manner. Crucial to Maxwell's method for predicting the weather was his revelation that all cycles of the same length turn at the same time, and that all cycles are related in one way or another to this basic energy cycle.

Professor Raymond Wheeler, who was head of the psychology department at the University of Kansas, invested over twenty years' time and the efforts of a staff of over two hundred to compile incredibly detailed records of three thousand years of weather and the apparent cycles which run through them. Nearly two million separate pieces of information about weather in history were entered on cards and supplemented with maps and charts.

Wheeler was able to chart types of governments, human achievements, wars, and shifts in cultural styles from one extreme to another and back again. He was able to isolate definite patterns of human behavior, as men and women reacted to climate changes.

He discovered a 100-year cycle, divided into four almost equal parts, that demonstrates that man has behaved differently—but *predictably*—during periods of warm-wet, warm-dry, cold-wet, and cold-dry weather. Wars, depressions, revolutions, cataclysmic events—together with tastes in architecture, musical expression, poetic meter, and the length of hemlines—have occurred at evenly spaced intervals. Wheeler's monumental research provides demonstrable bases for predicting what will happen in years ahead.

John Cejka, formerly a farmer of Laurens, Iowa, and now president of Cyclomatic Engineering, Inc., of Glenview, Illinois, began to follow the work of the old Weather Science Foundation, Crystal Lake, Illinois, in the early 1950s, when Dr. Raymond H. Wheeler was a faculty member of Babson Institute and the devoted astronomer Selby Maxwell had virtually forsaken all human company for the solitude of his observatory. The materials essential for such long-range forecasting were owned by the Weather Science Foundation. Predictions were made available on a limited subscription basis to industrialists, businessmen, farmers, and stock-market speculators, and had been used by agricultural and industrial leaders for more than thirty years. Cejka (pronounced *suh-jay-ḱa*) used the weather information, being marketed by Howard Baker, to enable the Cejka Brothers' farm complex to receive top-dollar prices for their crops and livestock.

The Cejka Brothers became involved with long-range weather forecasting after they were offered the service as a method of improving their agricultural operation. Cejka had very strong and positive instincts about Maxwell's theories, and he set about investigating the work of the Weather Science Foundation.

Howard Baker was at that time the president and the heart of the forecasting service. But the Weather Science Foundation was no longer as active as it had once been. Baker was in his eighties, and it appeared that the organization had simply grown old along with him.

Selby Maxwell and Raymond Wheeler were gone. There were a

couple of part-time meteorologists left to prepare the weather-forecasting charts; and Baker, serving as salesman as well as president, managed what had once been an elaborate service and a far-flung organization.

Cejka, a successful farmer and self-educated scientist, had the vision to realize what the Maxwell-Wheeler weather-energy cycle could mean once it was properly disseminated to the scientific community and the general public. He successfully tested the theories for himself and permitted certain trusted associates to do likewise. Then he suggested that they form a new corporation, buy out the entire weather science operation, and set about reorganizing its services.

In 1973 Cejka and several Midwest farmers and businessmen associates purchased the formulas to the Maxwell-Wheeler weather-energy cycle from Weather Science Foundation (Wheeler and Maxwell were deceased; only Baker survived) and began to structure an organization which would effectively disseminate the cyclic knowledge which had been known only to a select clientele list.

Years before, Cejka had studied the cycles to determine that Iowa farmland prices would reach an unprecedented high between January and March of 1974. Committed to an extensive period of development and research with the Maxwell-Wheeler material, John Cejka sold his farm at this time, then moved to Glenview, Illinois, to set about expanding the enormous amount of data and documents carefully preserved by the founders and staff in the offices and observatory of the Weather Science Foundation.

Cyclomatic Engineering's present system of forecasting was then rapidly developed.

Cejka and his associates named the first corporation the Orthic Science Corporation, and it became heir to all the documents of Wheeler, Maxwell, and the Weather Science Foundation. Later John Cejka coined the term "cyclomatics" to describe their activities, and they added a subsidiary to their organization named Cyclomatic Engineering, Inc., establishing a business office first in Glenview, then in Northfield, Illinois.

If you place the Wheeler curve over the Maxwell weather-energy cycle, you see that they match. If you then place the weather curve established by Dr. A. E. Douglass, who studied one million rings from fifteen-hundred-year-old Arizona yellow pines in order to derive a weather "diary" from the natural records of plant growth, you would find that this pattern matches the other two.

It is by combining the Wheeler-Maxwell curves that anyone can foretell the most desirable date to plan a picnic, plant potatoes, buy or sell stocks, or impregnate one's wife. Together, the Maxwell-Wheeler discoveries of weather-energy cycles and human ecology quite literally present us with a roadmap of time. We are not simply talking about when it will rain or snow—although in a time of impending famine, that in itself is vital information. Rather, the weather, the migratory patterns of birds, the breeding habits of mollusks, the waging of war, the price of commodities, the length of hair over the ears, every single enterprise and endeavor of man has its own cycle. Each of these cycles, in turn, is interrelated and affected by the basic energy cycle.

Once the astronomers at Cyclomatic Engineering have plotted the daily positions of the north and south air tides, they turn their maps over to their meteorological co-workers, who interpret them in the light of knowledge gained through extensive research into the weather cycles. Cyclomatic Engineering's forecast maps show the locations of the cold and warm and wet and dry air masses long in advance. Their forecast for each day is made one to five years ahead of that which the Weather Bureau has prognosticated.

"What do you offer a client of Cyclomatic Engineering?" I asked John Cejka one day as he was visiting my office in Decorah, Iowa.

"We provide him with monthly summary maps indicating the general temperature and rainfall for the entire United States. He has but to find his area on the maps and refer to the general weather conditions for his area.

"We also equip him with weather maps and instructions on how to use them on a daily basis, so that he can track the different air masses and precipitation areas which will be affecting him and his business throughout the year.

TEMPERATURE AND PRECIPITATION FORECASTS

In these charts, the broken lines represent the forecasted maximum temperature, minimum temperature, and precipitation for each day. The solid lines show the actual readings.

Figure 1

Month: January 1974 Section: Columbus, Ohio

Normal Average Temperature for Columbus in January is 31.2° F.

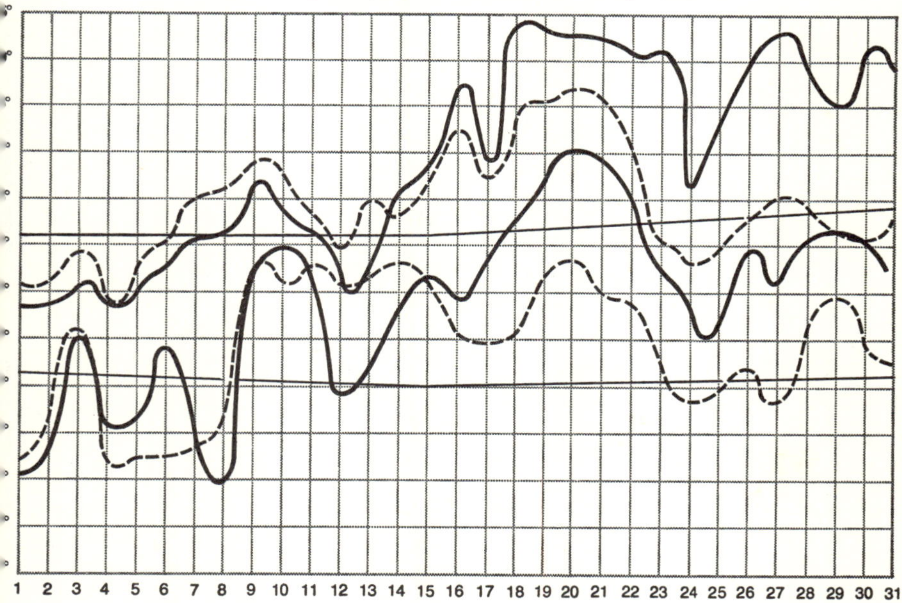

Normal Precipitation is 2.87 inches.

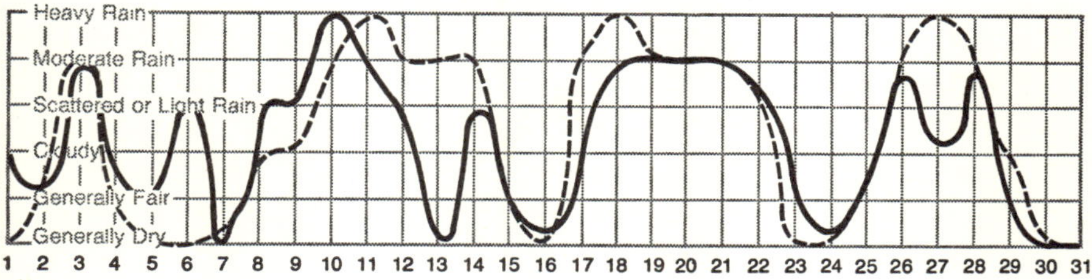

Figure 2

Month: February 1974 Section: Columbus, Ohio

Normal Average Temperature for Columbus in February is 30.3° F.

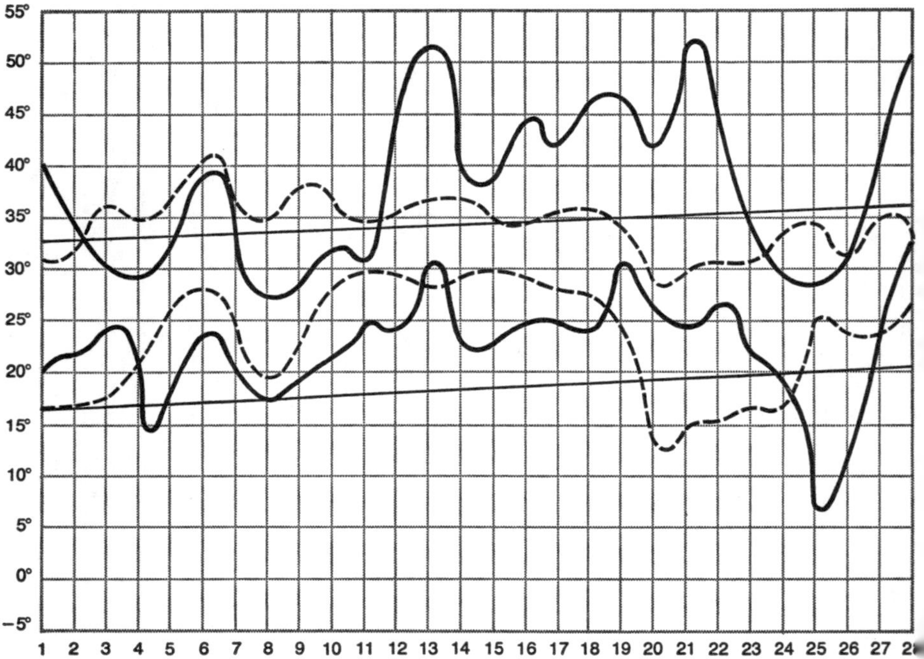

Normal Precipitation is 2.32 inches.

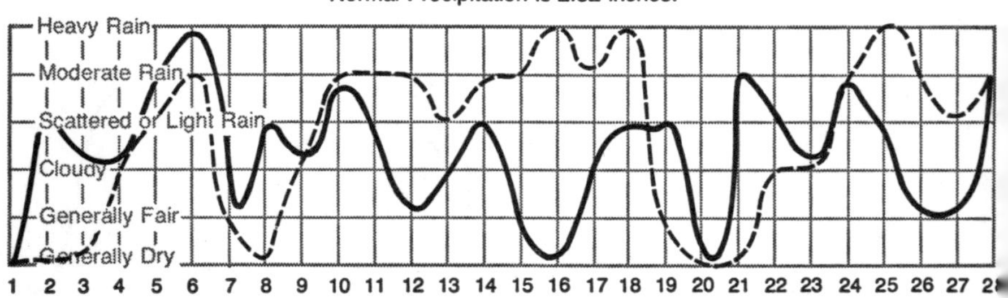

Figure 3

Month: March 1974 Section: Columbus, Ohio

Normal Average Temperature for Columbus in March is 39.2° F.

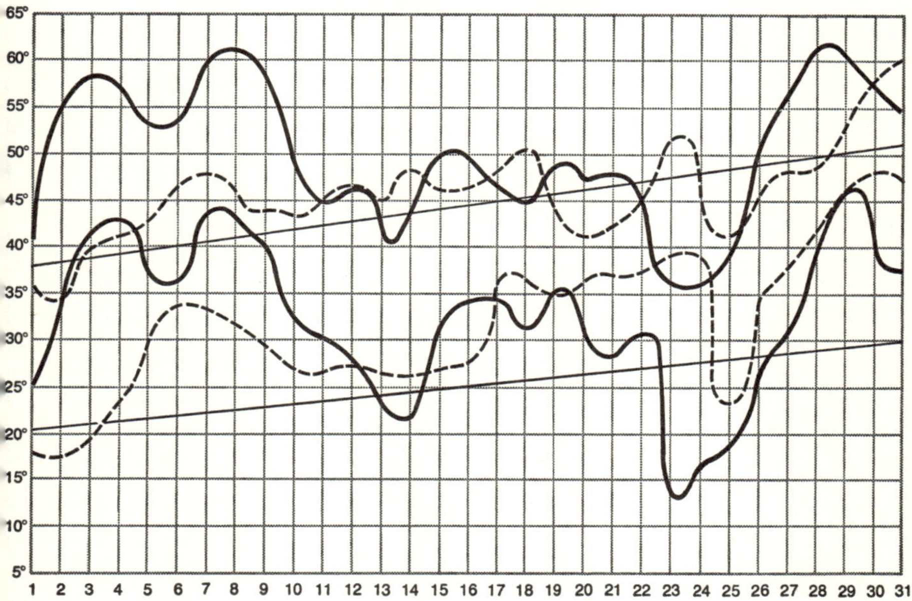

Normal Precipitation is 3.44 inches.

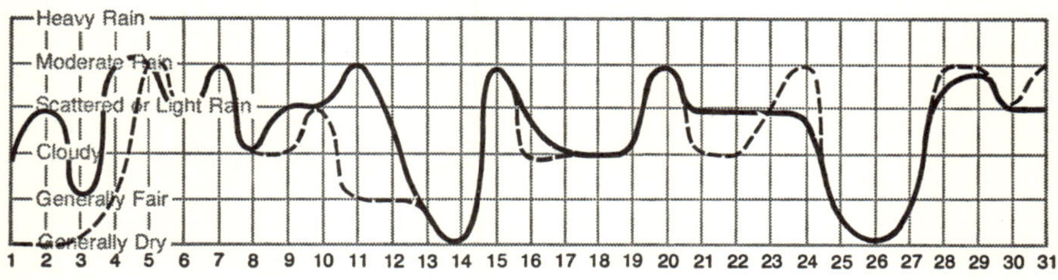

Figure 4

Month: April 1974 Section: Columbus, Ohio

Normal Average Temperature for Columbus in April is 51.2° F.

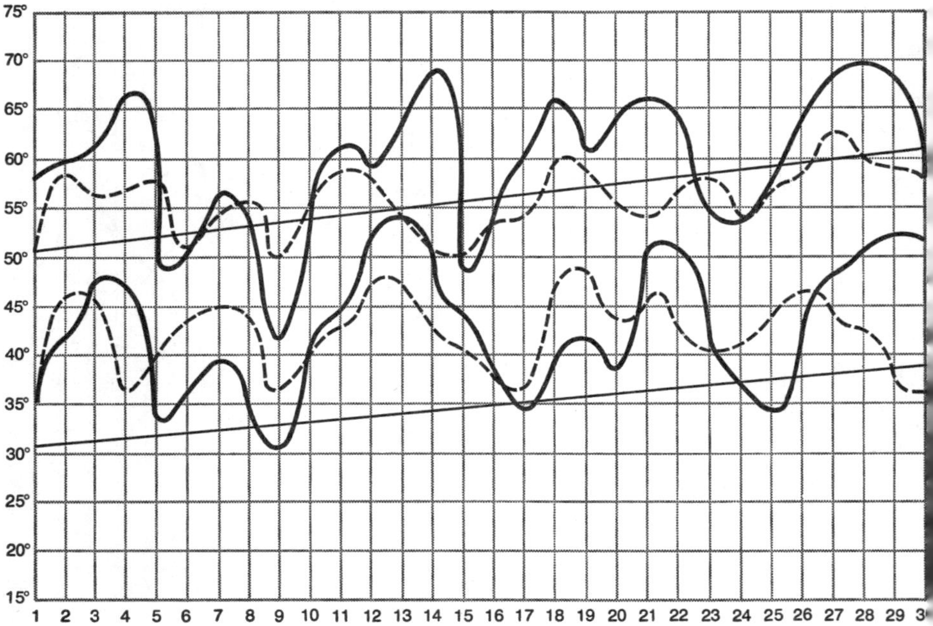

Normal Precipitation is 3.71 inches.

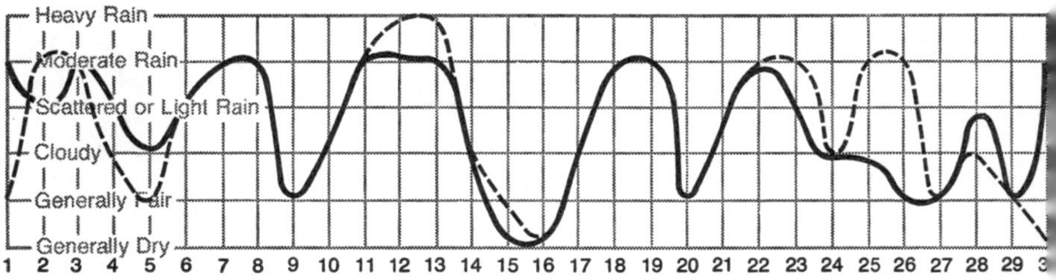

"In addition, there are weather summaries for commodity growing areas, and a tailor-made weather chart that specifically details the weather in his area and includes a daily high and a daily low temperature forecast. We also provide the services of our staff meteorologists on a Monday-through-Friday basis in case the client might have any questions regarding his interpretation of data."

Cyclomatics, Cejka explains, stands for automatic cycles and implies that weather cycles are not just due to happenstance—they are automatic and inflexible. They must occur as charted, just as the tides come in exactly as predicted. Cyclomatic Engineering, Inc., claims the astronomical calculations of the air tides are nearly 100 percent accurate. Their interpretations of the weather that will result in local areas, made from their basic forecast map, are about 80 percent accurate—but not because of errors in the basic calculations. For example, there may be a condensation area as indicated in the forecast, but showers can easily move around a small local area and miss a given farm in that region. Thus while the basic tide calculations are nearly 100 percent accurate in placing the cool, warm, wet, and dry areas on the map for each day, such local variations may sometimes reduce the accuracy of the forecasts to 80 percent.

Cyclomatic Engineering believes that an understanding of these weather influences results in important business advantages, because the weather trends can now be accurately forecast long in advance.

"The value of such advance knowledge should be obvious. When a farmer knows that it is going to be a dry June and July, he can plant accordingly and fertilize in a manner best designed to ride out the dry cycle. When a businessman knows the market will turn, he knows how long he can hold before he sells—or how long he will have to hold if he doesn't sell," Cejka said.

Weather research is of tremendous value to business today, because it can explain the present trends and their significance in world affairs—and the real causes of so many of today's problems.

But by recognizing that patterns of time do recur at rhythmic intervals, man can chart danger periods for his government. We can organize business, labor, and governmental agencies to produce and to distribute goods according to the best periods for growth. We can plan ahead for droughts and famines and store foodstuffs for the lean years. We can attempt to stockpile goods for periods of depression and inflation. And for the materialistically minded, the Maxwell-Wheeler curves also parallel—and thereby preview—stock-market fluctuations. Business cycles are integrated with the entire cultural pattern and, therefore, cannot fully be explained without reference to that pattern. Whatever factors set in motion the fluctuations of the culture pattern also trigger the general prosperity-depression pattern. Business cycles have been integrated with weather trends throughout history; they still are. The Maxwell weather-energy cycle and the Wheeler culture curve combine to provide an infallible predictor of the business cycles of all nations.

2

Selby Maxwell and the Science of Weather Forecasting

It is apparent that Selby Maxwell had been working on an autobiography over the course of many years. Throughout the voluminous notes he left in his observatory, one finds numerous sheets of graph paper on which his familiar printed prose details his life story.

What emerges from the carefully printed handwriting, is a portrait of a highly industrious and moral man. At the same time, it seems apparent that Maxwell considered himself to be set apart from his fellows, and at times his tone becomes a bit defensive. If one attribute stands out more clearly than any others, it would certainly be his single-minded dedication to his work and his conviction that his work would be of great practical use to humanity.

"Northwestern University," Maxwell writes, "was much like Evanston High, only more so, with a lot of gymnasium and athletics thrown in. Only running had any great appeal to me. My father did not look with a kind eye upon men striving together. He said life has enough strife and competing without going out creating some more. But running was useful. It could be used for getting to classes on time, for catching trains, and if the Civil War should start up again, getting away from Yankees or Rebels—or both.

"So I would have a deep interest in study, Father gave me a seven-inch telescope. It was fifty years old then, and it had a lot of things wrong with it. But I fixed it up, put it on an equatorial mounting, and made my very life revolve about it.

"Finally some of the neighbor boys, my own brother among them, smashed the lens. Here was another lesson of life that I got—namely, that the world, or some of it, is deeply resentful of any man who is superior to the common run. It was a very good lesson, and I never did forget it."

Selby Maxwell originally got into the weather-forecasting business because of a friendship with Gerald M. Stapleton, who worked for a brokerage firm at the time when Maxwell was science editor for the Chicago *Tribune*.

Maxwell tells how he resumed his friendship with Stapleton, a boyhood chum from his days as a newspaper carrier.

"He was working in the wire room of James E. Bennet and Company grain brokers; and he had just gotten a second, or night, job in the wire room of the Chicago *Tribune* (*circa* 1923). Mr. Stapleton was much impressed by my success in astronomy, and said he had something on his mind that he would tell me when the time was right."

A few days later, Stapleton told Maxwell that he had learned that Arthur Cutten, the reigning "Wheat King" on the Chicago commodities market, had secretly employed an astronomer, whom he was paying five thousand dollars a year to advise him on the wheat market. Maxwell was puzzled as to precisely what Mr. Cutten might be getting for his money.

"A Wheat King is a superbly successful trader in wheat and in other grains," Maxwell continued. "Some say that he is an unusually skilled gambler, but this is hardly the whole story. He is a speculator; that is to say, he buys grain in the hope of later selling it for more than he paid. . . . Needless to say, it requires much money to be a grain speculator. And above the money, it takes a keen grasp of world weather conditions to be a Wheat King —weather that has happened up to now . . . weather that is happening now . . . and most important of all, the weather that is *going to happen*.

"Speculation is carnivorous in the truest sense of the word. The spider that eats a fly, the cat that eats a bird, the Stone Age man who eats the flesh of another man—one and all are eating meat to get the energy that the meat contains. Money is another form of energy. It cannot be swallowed . . . but the 'money paper' can be taken to a bank and there exchanged for currency which will buy many a dinner.

"One of the essentials of speculation is that what one man gains, some other man must lose. The accounts always balance just this way. When one man becomes rich, another man, or more usually a group of men, must become poor. As might be supposed, this scheme of things does not make a successful speculator a much beloved man. In fact, the speculator is cordially hated, with not a little bit of envy and jealousy mixed in. . . . Every man jack of the crowds on Wall Street and La Salle Street would strip his fellows of their last dimes if only he could.

"The Street is filled with men who keep 'charts,' financial records of what has happened, and one and all try to predict the future. It is enough to be thrown out of some brokers' offices to so much as take a chart out of the pocket. Beneath the surface contempt for charts and chart players, there is a deep fear of charts.

"They say on the Street that of the people who come there, 98 percent go broke, 1 percent breaks even, and 1 percent makes money. Most novices at trading do not last long. No one willfully robs them: they cheat themselves. Professionals seldom outrightly lie to the novices, but they fill the air with half-truths—which in trading are often as bad as downright lies and harder to detect.

"No sooner does a man start trading than psychological laws go to work to alter his mind. A loss right at the start is often the best thing that can happen, for a thumping loss may knock the inner conceit into a cocked hat and thus save the man. To win opens the floodgates of conceit, and in place of the usual gamut of emotions —love, charity, idealism, kindness, fondness—there in time grow to be only two, fear and greed.

"When trading, thoughts of women, love, home, are positively poisonous; and they will cause losses. The successful trader keeps such thoughts in mental compartments where they cannot hurt him; and after a time, they wither. The successful trader gets to be a true miser. Unless fear and greed are rampant in his soul, he will never be successful at winning other men's money."

After a lengthy, mocking digression on assorted denizens of the world of speculative finance, Maxwell returns to the narrative thread of his story and to Gerald Stapleton.

Stapleton did some investigative work and learned that the Wheat King's private astronomer was also a professor at Northwestern University. Maxwell deduced that it was none other than his old professor of astronomy, Philip Fox, who was providing Arthur Cutten with the vital information which enabled him to make steady killings at the market. Although Maxwell never proved that Fox was the secret astronomer, his immediate conviction encouraged him to utilize his own knowledge to "beat" the stock market.

Maxwell was an astronomer, after all, and the two young men set out to duplicate Cutten's extraordinary success.

Stapleton and Maxwell rented a room on Superior Street, "amid the artists, poets, crackpots, and other odd gentry," where the activity of two "astrologers" would not arouse much curiosity or comment. Although they paid their rent promptly, their fellow lodgers were not so considerate of the landlady's own financial problems, so she had to bake pies to help pay her expenses. Maxwell and Stapleton bought several and ate them rapidly to keep them away from the numerous rats who shared their quarters. Maxwell became convinced that "eating pie is a distinct aid to the reasoning power." Years later when he was teaching college physics in the U.S. Signal Corps, he punctuated his lectures with things to eat.

Maxwell's previous experience with weather predicting had

been laborious and tedious. After the Versailles treaty had been signed, the *Tribune*'s editors realized that crop conditions in Europe would be a deciding factor in the new peace. Maxwell was assigned the task of plotting potential famine spots in Europe, and to record the information for a war atlas the newspaper was to release that year.

Upon completion of that assignment, Maxwell tried to solve the secrets of long-range weather forecasting. It had taken two years for him and his staff to test all the known systems in usage around the world and to declare them inadequate.

It was at this time that Maxwell reestablished his friendship with Gerald Stapleton and accepted the challenge of devising a formula that could accurately forecast the weather. Now the two men went through the *Nautical Almanac* page by page, trying not to carry any preconceived ideas of their own into the search.

"We were there to find out things," Maxwell writes, "not to prove how smart we were. As things turned out, none of our previously held ideas were the correct ones. We saved time and effort by booting out the old ideas right away.

"Always before, ideas of celestial influences had come from the old astrological thought of *position*. Nearly all the celestial mechanics taught in the university had, in one way or another, revolved around the ideas of *position*. Indeed, the whole *Nautical Almanac* is one vast collection of astronomical positions. The *Ephemeris* is a book that tells people where and when to aim a telescope. What we were seeking was the *time* to make a trade—a thing not immediately associated with right ascension and declination.

"We were, in fact, dealing with the third leg of the Tetrahedron of Space-Time-Energy, the third leg of a tripod upon which the Fourth-Dimension relations of matter in Space, Time, and Energy are sustained. It seems odd now, looking back, how completely foreign those fourth-dimensional ideas were in those days. Not many years before, in high-school geometry class, I had been laughed to scorn for barely mentioning the evolution of one conic curve into another via mutations of space."

Since childhood, both Stapleton and Maxwell had the habit of walking in solitude while thinking out a problem. Together they walked in silence along the shore of Lake Michigan and on Chicago's "crowded, but spiritually empty, pavements."

One day, Stapleton convinced his friend to enter a newly erected skyscraper and ride the high-speed elevators, like a small boy tempting a friend to join him on an amusement-park ride.

"They go so fast, they sicken you!" Stapleton complained.

"It was not the going fast that did the sickening," Maxwell wrote, "but the starting and the stopping. 'Stape,' said I, 'I think we have found what we have been looking for! It is the acceleration, plus or minus, that counts; and not at all where we are, either at the start or at the finish! This is Newton's law of motion, that everything goes on in a straight line *until* another force acts upon it. How very dumb we were not to see this right off.'

"Stapleton and I walked back to our studio room on Superior Street, and we went to work with the *Almanac* with feverish interest—now that we knew what to look for. No longer was the *Almanac* column after column of figures of position in right ascension and declination. Now we saw that it was *how* something went from Alpha to Alpha Prime that counted. It was the *going* from one position to another position that counted, and not the *being* in either the first or the second position."

Maxwell explained that they were facing exactly the same problem which people had confronted when making dynamos. How big or how small the magnet was did not make any difference. As long as the loop of wire stayed put, nothing happened. But the instant the wire loop *moved* in the magnetic field, a current of electricity shot through the wire. When the motion stopped, the electric current also stopped.

"When we go from one astronomical position to another position," Maxwell theorized, "we get a surge of market energy—better called Weather Energy, I think. As soon as there is no more astronomical motion, there is no more Weather Energy, or motive power for market action.

"A dynamo is an electromagnetic machine, but weather and markets are a gravity machine! They do not constitute a machine of wheels, wires, and armatures, but a machine of analogues or mechanical pictures.

"There is, of course, only one actual gravity machine—the Earth, Sun, and Moon. What we can make are little mechanical pictures of the working of this or that part of the Earth-Moon-Sun machine."

With one man reading the *Almanac* figures and the other man operating a desk calculator, they computed the accelerated motions of every item in the *Ephemeris*.

In another section of Maxwell's notes, we find the materials which they assembled for their chartings and computations:

Market Weather-Energy Computations

Object: To find three major energy curves which when applied together to Longus computations will give the Long Trend of Weather Energy plus numerous intermediate movements as applies to *market actions* within one day of occurrence.

Materials needed:

1. The *American Ephemeris* and *Nautical Almanac*
   a. Sun (heading)
      (1.) Var. of var. of apparent right ascension.
      (2.) Var. of var. of apparent declination.
      (3.) Var. of var. of radius vector of the Earth
   b. Moon (heading)
      (1.) Phases of the Moon
      (2.) Longitudes to obtain 1st of Aries to 1st of Aries to obtain date spacing on base lines.
      (3.) Var. of var. of transit of the Moon.
2. Graph paper, preferably ten squares to inch.

Notes

1. Dates used will be astro dates, throughout construction of major energy curves.
   a. These curves will be denoted A, B, & C.
2. Base lines will be laid out using spacing obtained from Aries to Aries with standard Aries month being 8 9/32 inches.

3. Spacing for all values of var. of var. plotting will be 3/10 inch on three squares of the graph (i.e., 3/10″ = value 1; 6/10″ = value 2).
   a. There is a template, but for all practical purposes this scale is close enough.
4. Aries to Aries dates will be obtained from the longitudes of the moon. When going from 359° to 1°, figure for 360° to nearest. 1/4 day.

With the assistance of Stapleton, Maxwell developed the fourth-dimensional mathematics which enabled him to discover the Weather-Energy Cycle. "We found," Maxwell discloses, "that it is the *second* differences between consecutive numbers that count. These second differences in various combinations, and lagged with the enormous hysteresis of so many days, give us the place to look for the motive power that makes the markets and the weather go. I have had many men scoff at my discoveries. Not one that I ever met had gumption enough to understand hysteresis, or lag.

"Once I was walking along a railroad track where some men were driving spikes a half a mile away. First I would *see* the hammer fall, then I would *hear* the blow through the iron rails—then I would hear the blow through the *air*. It was all the same action, but recorded with three lags. Most of my detractors have seen men working on railroad tracks, but they did not possess enough sense to connect *two* noises with *one* hammer blow."

In order to test their theory, Stapleton and Maxwell brought together some money and invested it in wheat futures. "We did not aim for a killing, but we did want to own automobiles. We made money right from the start."

Successful in their stock-market speculations, the two friends began to make plans to establish a market-forecasting service.

They might have gone on for years, making just enough money off their hypothesis to keep them both in comfort, but then tragedy struck.

"Stape backed his car into a freight train's path," Maxwell wrote beside the word "EXIT" he block-lettered on his graph paper manuscript. "He and all his family were killed. Most of them must have died very quickly, for the train hit so hard that it cut the car almost in two."

A sorrowful Maxwell lost all interest in setting up a company.

"I went on alone, but this grim tragedy never left me. I made a few fair trades, but I did not have the heart to found a trading firm, as we had planned. Instead, I played in astronomy and weather, perhaps, subconsciously, waiting for my grandsons to grow up." Maxwell virtually lived inside his own observatory and recorded his findings in maps, formulas, equations, charts, and scrolls. He often spent eighteen hours a day in his observatory, perfecting means of utilizing the weather-energy cycle and working on other inventions.

According to the theories of such men as Selby Maxwell, long-range weather forecasting is an exact science. Systems of weather prediction utilized by the Weather Bureau are based on observation, on information from reporting stations, and on the sciences of meteorology and astronomy. But "Selby Maxwell," John Cejka told me, "discovered the primary weather-energy cycle that governs all weather—past, present, and future. All cycles of the same length turn at the same time, and all cycles are related in some way to the Maxwell cycle."

"Can you state just exactly what the Maxwell cycle is?" I asked.

"It is based on the movements of the Earth, Moon, and Sun—but chiefly on the interaction of the Moon and the Earth. Maxwell stressed that as the Moon and Earth revolve around one another, there is a differential in the pull of gravity. Since the Moon orbits Earth in an ellipse, there must be a period when it travels at a faster speed, and another time when it slows down. Everything on Earth is constantly subjected to two forces—the gravitational forces and the centrifugal forces from the spin of the Earth on its axis."

Ever since our early school days we have been told about tides and the Moon. The key to an understanding of tides lies in the law of gravitation.

The Sun, most massive of all bodies in our solar system, holds Earth in an orbit. The Moon is about 1/81 times as massive as Earth, but because it is so much closer to Earth than is the Sun (250,000 miles, opposed to 92 million miles), it is held in tight gravitational bondage to Earth. Part of the payment Earth makes for this bondage is the deformation it suffers as the result of the tides.

When the Moon is aligned with Sun and Earth at New Moon or Full Moon, the tides are greater, because the Sun and the Moon cooperate in raising them (spring or flood tides). If, however, the Moon and the Sun are at right angles to Earth, as at first or third quarters, then the two are opposed, thus raising smaller tides (neap tides).

The Moon rotates about Earth, taking a period with respect to the Sun of about 29 1/2 days. Since Earth is also rotating, one might suppose that each location on the planet should have two high tides and two low tides each day. This is not so for several reasons.

The Moon moves in a plane more in line with the Earth's orbital plane than with the equator. Thus, in effect, Earth's axis is tilted toward the Moon. Thus at times the peak of the tidal bulge is almost 29° north of the equator, or nearly a third of the way toward the pole. Farther north than this, only one large and one small tide are observed each day.

Earth is not a smooth ball. It is rough and flecked with great ocean depths and continents. The coastlines are irregular and cannot be easily defined by practicable mathematics. The waters possess natural wave motions which tend to interfere with the tidal movements. The bottom of the ocean has a frictional drag on the water. The result of these factors, Selby Maxwell found, is that theoretical calculations must be modified by intense observation of past tidal events.

The Moon comes closer to Earth (perigee), goes farther away to its furthermost position (apogee), than comes nearer again—an irregular orbit producing a 20 percent change in tidal events. The *Nautical Almanac* lists positions of the Moon so that in cases where the port tide characteristics are not known, a good qualitative calculation can be made of high, low, neap, apogee-neap tides, and so forth.

"Now, scientists learned long ago how to calibrate tides at a given time anywhere in the world," John Cejka told me. "This is not really too difficult, because we know the Moon and the Sun affect

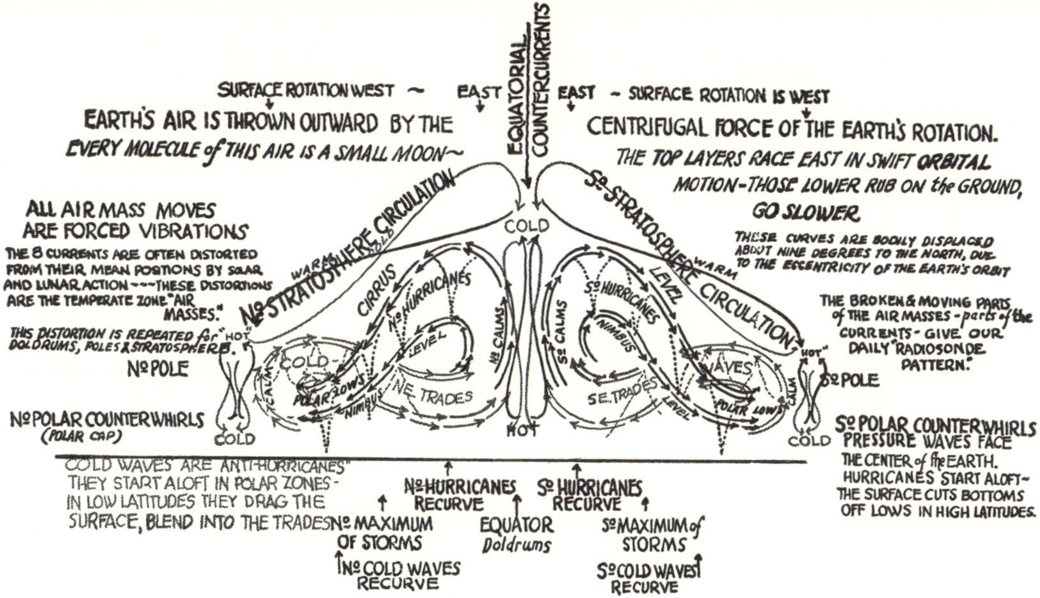

DIAGRAM OF THE AIR MASS CIRCULATION OF THE EARTH'S AIR

EXPLANATION — DIFFERENCES IN ORBITAL MOTION OF THE EARTH GENERATE FORCES THAT WOULD TEAR THE EARTH TO PIECES WERE ITS GRAVITY WEAKER. ONLY THE STRONG GRAVITY OF THE EARTH HOLDS IT TOGETHER AGAINST THESE DISRUPTIVE FORCES. ALL AIR MASSES ARE PROJECTILES, THEY MOVE AS PROJECTILES. EACH EARTH MOVE SETS THE WHOLE QUARKHA OF EACH PROJECTILE IN MOTION IN ALL PARTS. NO VACUUM OCCURS IN AIR, BUT MOVEMENT SWIFTER THAN NORMAL LOWER THE PRESSURE OF THE AIR, WHILE SLOWER MOVEMENT RAISE IT, AS IN JETS.

A QUARKHA IS A SPHERICAL CURVE of 720°.

THE AIR of EACH of the 8 GREAT WIND SYSTEMS of the EARTH IS A QUARKHA, EACH WITH 3 COMPANIONS - at POLE, at EQUATOR - over HEAD.~ ALL WHIRL WITH the WHIRL of the EARTH - WITH A CONSTANT of MOMENTUM.

A + or − ACCELERATION from the MOON SPEEDS or RETARDS THEM; THESE VARIATIONS ARE "AIR MASS" RESIDUALS

water through a gravitational pull. But scientists have learned that the causes of the north and south air mass movements, which cause the cool- and warm-weather waves, are north and south tangential "tidal thrusts" timed by the movements of the Moon. These tidal thrusts are the force which starts the air masses in motion. Since it is the librations of the Moon which cause the tides, astronomical mathematics can calculate these movements for every day a year in advance.

"Such a systematic method of measurement forms the basis of Selby Maxwell's forecasting system, but his greatest contribution was to delineate the correct time lags which cause the turbulent upper air masses to act in the predetermined manner which they do. Maxwell originally undertook this method of long-range forecasting on the theory that if north and south tangential tidal thrusts were strong enough to vary the radial tides, they might also initiate the movement of air masses. When the daily locations of these tidal thrusts were plotted, they revealed close correlation with the daily location and movements of the air masses as shown on Weather Bureau observation maps. He learned how to plot the air masses over landmasses, and this is the key to predicting the weather.

"Air tides are influenced by the Moon just as sea tides are. You see, in essence, that jet stream in our upper atmosphere is equivalent to another Moon, another satellite of the Earth. The jet stream travels at velocities up to six hundred miles per hour, faster from west to east than the Earth is turning at the equator. Hence, these air tides are in orbit, and the orbit of one satellite affects the orbit of the other.

"Maxwell developed his formula of weather prediction by being able to chart the movement of the jet stream by the motion and declination of the Moon. The proper determination of *Time,* rather than *position,* of Earth, Moon, Sun, and upper air masses, led to Maxwell's perfecting the formula which has permitted researchers privy to this information to accumulate a thirty-year record of accurate weather forecasts made one to five years in advance. The center of the jet stream is always where the storm activity is. It comes down during the warmer seasons, and it shifts back up during the winter months because of the inclination of Earth. But the jet stream is in free orbit. And we know the cycles because we know the tides of the jet stream. This makes it possible for us to predict major storms, not just three or four days in advance, but three or four years in advance. Using Maxwell's research, we can predict the movement of air currents as accurately as science can predict the height of tides."

Eventually, Maxwell's work was combined with that of Raymond Wheeler and incorporated into the interpretive weather service offered by Howard Baker. But most of Maxwell's data has never been made known to the scientific community.

After he was partially paralyzed by a stroke, Maxwell constructed a special motorized elevator alongside his observatory so that he might continue in his experiments, which he did until his death on March 3, 1971.

The last section of Selby Maxwell's manuscript that I have been able to locate concludes with these words:

"We are entering an age of space and computers, with more and more of the future being revealed to us. Market action is one of the places where the future is very easily revealed. Only ignorant people will refuse to see this."

3

Raymond Wheeler and the Weather Gestalt

Raymond Wheeler first became interested in world climate because of a problem he encountered while teaching the history of psychology at the University of Kansas. In his own words:

"Since 1912, developments in psychology have proceeded along two theoretically opposite and incompatible lines. On the one hand there have been behaviorism and conditioned response psychology that have followed the strictly orthodox, mechanistic principles of association. These psychologies assumed that human behavior, regarded either subjectively or objectively, was a complex pattern of phenomena *built up* by bond-forming processes from simple elemental responses, in such a way that, piece by piece, the resulting pattern of the whole was obtained mechanically from preexisting parts. The basic fact was the simple experience. Mind was passive.

"Opposed to this view there sprang up several varieties of gestalt or organic psychology whose common thesis is that behavior, from the beginning, is an *already integrated pattern* of responses, however simple: that complexity is achieved not by the procedure of obtaining wholes from parts, but of parts from wholes, through processes known as expansion and individuation (differentiation). The basic fact is the whole mind, the personality. Mind is active.

"Heinrich De Bary, an eminent nineteenth-century botanist, expressed the issue succinctly in 1868 when he said, 'Cells do not make the plant; the plant makes the cells.' So interesting a dilemma could not help but stimulate a historical investigation. The issue was fundamental. The two modes of approach were both methodologically and logically opposite. Why should they ever have arisen?

" . . . a review of history yielded a surprise. The roots of mechanistic behaviorism and association psychology alternated with those of the organic points of view, and the histories of biology and philosophy paralleled the history of psychology exactly, in analogous alternations.

"Extending the inquiry further brought to light alternations in an entire culture pattern of which those in the three subjects first investigated were only special cases; i.e., they were merely parts of an integrated, fluctuating whole, involving all the other sciences, and in addition, art, literature, social and political trends—in fact all possible expressions of human energy. This fluctuating culture pattern turned out to be synchronized with the rise and fall of governments, nations, and empires—the organic expression coinciding with the rise of strong states and totalitarianism; the mechanistic with the weakening or fall of absolutistic governments and the growth of democratic institutions.

"This shifting back and forth carried with it reversals in attitude, feeling, judgment, purpose, degrees of tolerance and intolerance, et cetera, to such an extent that *human lives, on a wholesale basis, were free or not, happy or not, hurled into war or not, in accordance with these fluctuations. For both international and civil war were obviously dominated by these changes in the human behavior pattern.* The former were closely associated with organic, absolutistic, and totalitarian cultures, while the latter were associated with atomistic, mechanistic, utilitarian, and democratic cultures.

"A curve of these culture 'cycles' was plotted before any hint of their relation to environment had been discovered or even suspected."

One evening, Raymond Wheeler presented these facts to a group of scientists. During the discussion that followed, a member of the group inquired if the cultural fluctuations had been compared with the tree-growth curve obtained by measuring the annual growth rings of the 3,300-year-old California sequoias.

"The question was at first puzzling," Wheeler confesses in his journals, "for what possible relationship could there be between . . . cultural fluctuations in human society and . . . fluctuations in the rate of growth of the big trees of California? At any rate, the inquirer who had recently seen the sequoia curve thought he recognized a similarity between it and the culture curve, which had just been presented.

"I had never . . . even heard of a tree-growth curve. But take a look at a freshly cut log, and you will see a series of concentric rings beginning in the center and extending out to the bark of the tree. Look . . . closely, and it will be noticed that they vary in width.

"In those geographical localities where seasons are regular and stable, trees lay down one layer of light cells during the winter. An expert can measure these rings and plot the measurements in the form of a curve. The widest rings are laid down during the longest growing seasons; i.e., during those years in which there are long seasons favorable for rapid growth. It turns out that these are the years when the weather trend in the vicinity where the trees are located was warm and wet. In a warm year the growing season is longer. Warmth, moisture, and the long growing season, all taken together, produce the widest rings. On the other hand, a narrow ring is produced either by a very dry or short growing season or a combination of the two. It so happens that the longest periods when the trees were growing slowly are times when the years were both cold and dry . . . winters long and rainfall poor throughout the year.

Figure 6 TREE GROWTH IN RELATION TO THE CLIMATIC CYCLE (SCHEMATIC)

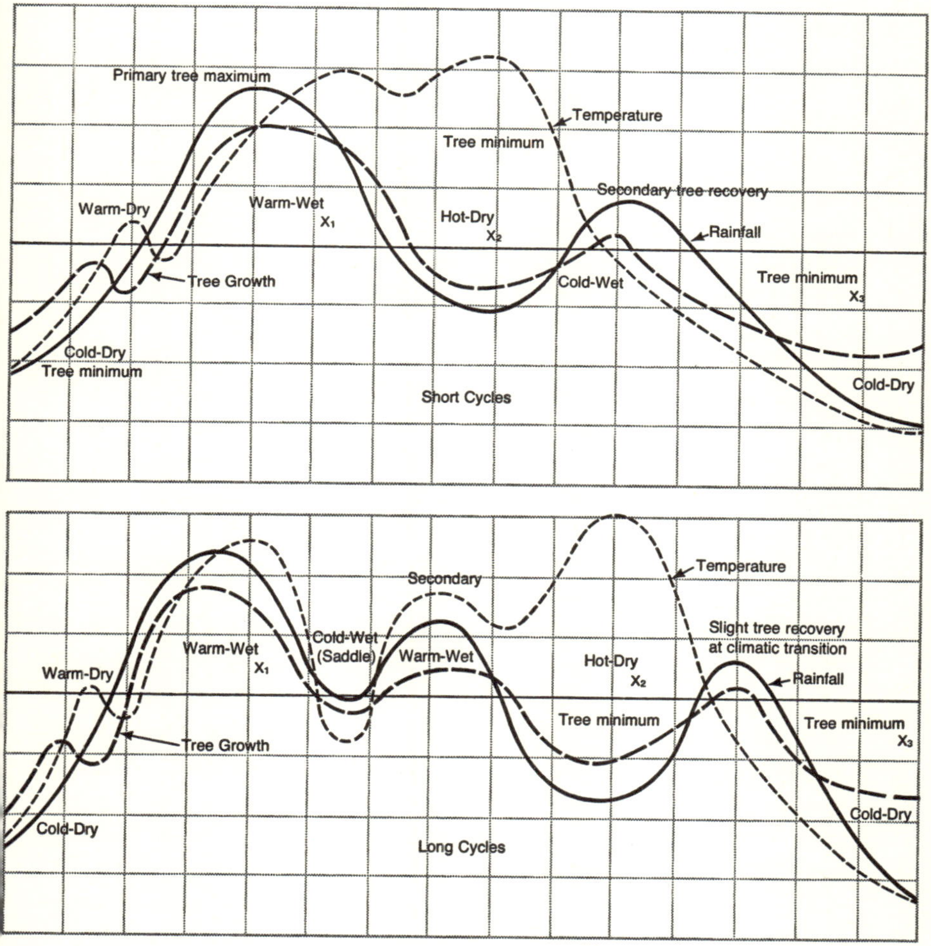

"Studies show that the sequoia tree-ring curves indicate what the world weather trends were like *much of the time* down through history.

"The peaks and troughs of . . . the sequoia curve and the culture curve corresponded so well down through the centuries that . . . laws of chance could not account for the similarity. There seemed to be only one plausible explanation, namely, that . . . fluctuations in human behavior were associated with fluctuations in world weather trends, or climatic fluctuations. In other words, . . . this led to the suspicion that man's behavior might possibly be conditioned by climate in a way hitherto unsuspected.

"The next questions were: Did the fluctuations in tree growth adequately reflect the weather trends? Were these weather trends really worldwide in scope? Accordingly, an intensive study of human behavior in relation to climatic changes was then begun.

"From there on evidence was collected as rapidly as possible. . . ."

Tree-ring curves are only one of many sources of information which Wheeler used to check his history of world weather. There were measured temperature and rainfall from weather stations widely scattered over the world. However, effective government weather bureaus were first established in the early 1870s, about one hundred years ago. At best, temperature and rainfall measurements go back only two hundred years, and only a very few records extend back that far. Again, quoting from Wheeler's notes:

"After measured temperature and rainfall records for the world as a whole were investigated, a study was made to ascertain whether various kinds of miscellaneous evidence regarding the weather could be used, when pooled in sufficient amounts, to locate successive phases of the weather cycle. In newspapers, together with the reports about weather events found in the annual chronologies of the *World Almanac*, it was found that reports of severe droughts and heat waves clustered during times when weather station records demonstrated . . . warm-dry years; that reports of extreme rainfall, storms, and floods clustered during . . . climaxes in world rainfall; that reports of excessive cold and snow clustered during . . . cold-wet years; and that reports of cold with accompanying droughts clustered during years known to be cold and dry. Thus it turned out that miscellaneous evidence, assembled in sufficient quantities, could be used to help locate . . . even the shorter cycles. This evidence would not tell us . . . the average temperature or average rainfall in any given locality, but it would tell us when the different phases of the cycle occurred.

"Accordingly, a search was made for the evidence. The types of available evidence . . . are as follows:

> 1. Chronologies kept by various governments, army headquarters, monasteries, or other interested organizations or individuals, containing items about the weather such as droughts, heat waves, cold waves, excessive rain, floods, storms, excessive snow, famines, crop failures, locust plagues, forest fires, freezing and thawing dates, harvesting dates, lake and river levels, glacial retreats and advances, travel in the mountains (passes open for lack of snow or passes closed by excessive snow), the freezing over of large bodies of water, the freezing over of the Baltic Sea between Germany and Scandinavia, the blocking of coastal commerce by excessive ice, and so on. Thousands of individual items can be found if the search is long and intensive enough.
>
> 2. Lake and river levels as determined by geographical and geological analysis. This includes the location of old beaches on hillsides and dating them as closely as possible, and buildings now under water.
>
> 3. Tree rings.
>
> 4. Reports of sunspots large enough to see with the naked eye, and of aurora. Reports of sunspots go back almost to the time of Christ. When a group of large sunspots is seen, it is certain to turn cold for a period of years soon thereafter, and large spots are likely to recur during long cold periods. Warm periods are noted for their absence of large spots.
>
> 5. Pollen analysis of successive layers of the soil indicating changes in the types of forests.

"A special study was made of this miscellaneous information from the past in its relation to the story as told by the longtime

Figure 7 EUROPEAN CLIMATE, 1740—1930

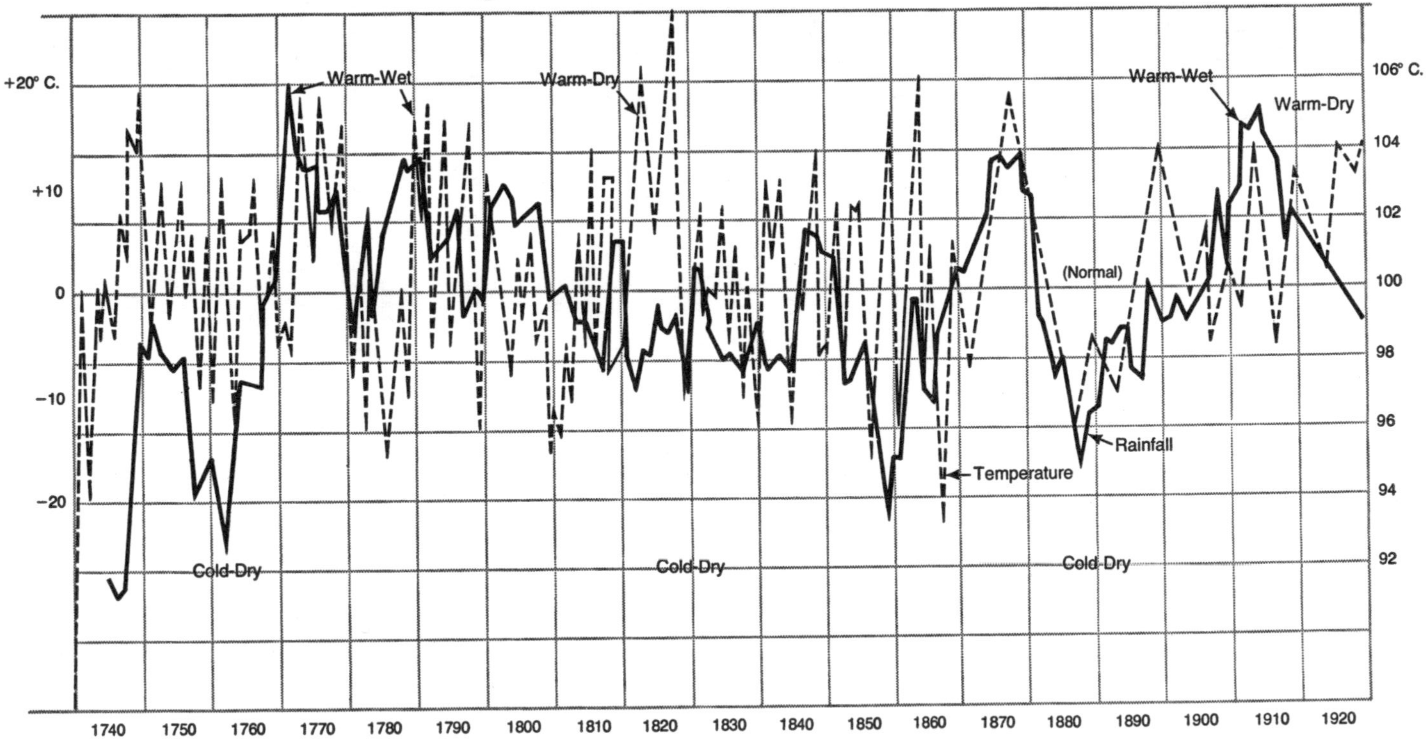

tree-ring curves. The story as told by the tree rings, taken as a whole, and the masses of miscellaneous evidence agreed with one another to such an extent that it was found possible to deduce from the sequoia tree-ring curves the different phases of a 100-year cycle. These deductions were confirmed sufficiently by the miscellaneous evidence to justify a reliance on the sequoias at least 80 percent of the time when miscellaneous items were scarce."

"Obviously others were working on weather problems and weather-energy cycles before Maxwell and Wheeler," I pointed out to John Cejka.

"Sure they were," he said, his words coming in short bursts synchronized with puffs of smoke as he lighted his pipe.

"Like so many other great innovators in man's own cyclic evolution, these two men just tuned in to ancient truths that had been lying about unnoticed for centuries," Cejka said, laughing.

John has a habit of punctuating his conversations with regular staccato bursts of laughter. The listener is never being mocked for his lack of knowledge, however. Spontaneous laughter is simply another means by which John Cejka expresses his enthusiasm.

"Ancient Hindu literature is full of references to longtime cycles of man's supposed rise and fall," Cejka said. "Babylonian and Chinese literature contain similar references. A lot of the ancient Greeks held to a rhythmic conception of history, and they were also conscious of a possible relationship between historical events and climate.

"Cyclic theories of history were popular during certain periods in the Middle Ages, in the sixteenth and seventeenth centuries, and again during the early nineteenth. Geographer Ellsworth Huntington correlated the rise and fall of civilizations with climatic changes.

"Huntington believed that history is, in part, determined by the simple process of climatic change from warm to cold and wet to dry, and also by the degree to which the environment is climatically stimulating. According to Huntington, the ancient civilizations were in the southern part of the temperate zone, because the Earth was cooler then. Northern Europe and central Asia were too cold, probably too dry, when Greece had its glory and Rome had its grandeur. Huntington's work maintains that the main storm tracks of the north temperate zone were then farther south than now. When they moved north, along with an expansion of the temperate zone, civilization moved north along with them.

"During the twentieth century, with a great interest in patterns of human behavior, an emphasis returned on establishing a relationship between weather-energy cycles and all phases of human activity.

"Now, of course, a study of the literature on periodicities in rainfall, temperature, tree growth, volume of business, and so forth can certainly leave you pretty confused," John chuckled. "And some of the historical works on alleged secular cycles and recurrent periods of activity simply can't survive critical analysis. So it is with certain studies which purport to have found definite secular periods in which climatic and historical events repeat themselves in synchronized fashion."

"Then what you are saying, John," I tried to summarize, "is that any effort to visualize the history of climate *solely* in terms of single, secular periods—or to regard culture cycles in the same manner—hopelessly oversimplifies the problem."

But I wanted to ask John one last question before we took a break for lunch: "How do you know that Maxwell's and Wheeler's own interest in the weather-energy problem didn't bias them to the extent that they saw facts that were not so? How do you know that they did not select the facts that seemingly proved their theories and left out facts which would disprove them?"

I certainly had not wanted to appear as a philistine in the midst of a room piled high with the research of two dedicated scientists, but I did want to hear how John would answer such a question.

"Well, first of all, Brad, I've proved these theories in practical application dozens of times over, so I know these men were correct," John began, "but let me say this for Wheeler and Maxwell: They began their research to gather all of the facts. Their opinions and conclusions were not preconceived. The facts were not studied to prove a point. Their research was begun for the purpose of learning why there were complete reversals in viewpoints, in governments, in the sciences, and in all human affairs from time to time. Their research was begun to build curves of those trends in human affairs.

"It was only later, when these trends were compared with the curves of the weather trends and cycles, that the correlation was discovered. They were not previously trained in those conclusions. Therefore, there was no searching for facts which would prove a conclusion previously accepted.

"Only after the trends and cycles in human affairs were found to be in direct correlation with the changes and trends in the weather was the research actually begun to collect all of the facts in relation to both. The purpose was to learn if the correlation would continue *after* all of the facts were assembled.

"All branches of human affairs and endeavors and achievements, arts and sciences, and so forth were collected. Twenty thousand pieces of art were studied. Literature throughout history was analyzed. More than eighteen thousand battles and military campaigns were studied, and the facts collected and assembled in relation to their timing.

"Moreover," John emphasized with several vigorous thrusts of that forefinger, "Maxwell and Wheeler didn't list historical data from only a few parts of the world. They ploughed through thousands of volumes of history relating to every known nation that ever existed. The Orient as well as the Western world was carefully analyzed.

"There can be no selection to prove a theory when dedicated individuals go through the trouble to get all the available facts possible. Every phase of the Maxwell-Wheeler research has been done with a corresponding thoroughness. They used all the facts they could get their hands on.

"Only then it was found that the timing of the international wars and civil wars, the timing of democratic and totalitarian governments, and so forth were found to be in direct correlation with trends and cycles in the weather. The conclusions were developed from a study of these correlations. The correlations or the facts were not studied later to prove the conclusions.

"In all cases, the two sets of data were plotted separately and by accident found to agree. There could not possibly have been any selecting of favorable facts in advance. All of the facts were assembled first."

"Maxwell and Wheeler obviously spent an enormous amount of time checking and cross-checking references before they made their own predictions," I said. "They both seem to have been very conscientious and cautious men."

Cejka laughed. "You could not believe the amount of hours these men and their staffs put in on these problems. Before any sense can be made out of the problem of periodicity, either in climate or in history, extremely long and extremely complex curves must be exhaustively studied for their mathematical properties as wholes—that is, for accelerations, decelerations, hidden or potential periodicities—those that have vanished for a time, but are undoubtedly, somewhere, potentially present in the causal matrix—patterns, volleys, and so on. These complex curves have got to be even more than the sum of their parts."

"And then when these curves have been carefully assembled," I put forward, "you feel that there is not a single aspect of human life which escapes being influenced in important ways by climate."

"It is impossible to read the vast amount of literature on the problem of climate and human behavior, then to study Maxwell and Wheeler's exhaustive research and to become involved in the study at first hand, without coming to such a conclusion.

"This does not mean that they solved all the problems pertaining to the relationship of weather trends to human reactions," John admitted, "but their answers demonstrated the thoroughness with which they attempted to do their work. We at Cyclomatic Engineering are using their high standards as our guidelines."

4

Charting the Cycles of World Weather

More than four hundred researchers worked for over twenty years on this study of the influences of weather cycles on the affairs of mankind. At Dr. Wheeler's instructions, they examined the influence of weather cycles upon all basic human affairs for about three thousand years of history. They discovered that changing trends in government, in political affairs, in wars, in culture, in all sciences, were in direct relation to the world weather trends and cycles.

"Maxwell and Wheeler ran into an important 100-year period, both in history and in climate," Cejka told me. "This same interval had been noticed by a German researcher named Mewes near the end of the last century."

"Eventually our questions were answered in the affirmative," Wheeler writes. "Tree rings in sufficient quantity locate the majority of the warm and cold and wet and dry phases of the 100-year cycles in the weather trends; the weather trends themselves are generally worldwide in scope."

Wheeler summarized all of the climatic evidence in the form of a climate curve, showing the positions of the phases of the 100-year cycle with their interruptions. This curve was then compared with the succession of cultural events. It was evident that the climatic and cultural fluctuations had occurred together all down through history. Either the cultural cycles were conditioned by the climatic cycles, or both were the results of the same unknown cause.

A study of available temperature and rainfall records pointed out that this 100-year weather trend or climatic cycle, worldwide in scope, passes through a definite succession of stages, repeated in the same order as the cycle repeats. Additional evidence confirmed the existence of this cycle. In fact, Wheeler found it is possible to trace it back at least to 600 B.C.

"These are not mechanically or mathematically precise periods," John Cejka reminded me. "Sometimes that longtime weather trend may shrink or expand a bit. But Maxwell and Wheeler found that it averaged 100 years in length and that it has a definite pattern which tends to repeat itself."

The cycle visible in the sequoia tree-ring curve is not always 100 years in length. It shrinks at times to 70 years and lengthens at times to 120 years. It is usually either 90 or 110 years long. Dr. Wheeler was able to define that this very distinct 100-year world weather rhythm, or cycle, was made up of four basic phases—warm-wet, followed by warm-dry, then by cold-wet, then by cold-dry.

"The intervals forty-five, fifty, and fifty-five have also been found to be common," John Cejka said. "Another German, Zasse, and an Englishman, Wright, along with several others, have referred to a periodicity of fifty years relative to wars." And indeed, rainfall and temperature measurements for the last two hundred years also point to rhythms within rhythms, or cycles within cycles. The warm half of the 100-year cycle, approximately 50 years in length, is interrupted every so often by temporary drops in temperature. Usually these drops in temperature occur about every 11.2

years, at sunspot maxima. Sometimes the interruption will last for only a year or so, sometimes for 10 years.

The cold half of the 100-year cycle undergoes corresponding interruptions. Every so often, temperatures rise for a year or two, or up to a period of 10 years. Usually this temperature rise centers around a sunspot minimum.

The wettest times on the climatic cycle occur during transitions from the long cold phase to the warm phase and back again. The same type of event occurs in the annual weather cycle of the seasons. Over large areas of Earth, the wettest times of the year are the spring and fall, as it is turning warm or cold. In fact, the annual cycle reveals the same successive phases as does the long 100-year cycle, namely, a cold-dry mid or late winter, a warm-wet spring and early summer, a warm-dry mid or late summer, a cold-wet fall and early winter, then a cold-dry mid or late winter again.

It happens, therefore, that the first part of the warm phase of the 100-year cycle is the wettest (this is the warm-wet phase) and the last part the driest (the warm-dry phase). Similarly, the first part of the cold phase of the 100-year cycle is the wettest (the cold-wet phase) and the last part the driest (the cold-dry phase). Temperatures reach their highest point in the longtime cycle during the warm-dry phase and their lowest point during the cold-dry phase. In other words, temperatures are more moderate during periods of normal or abundant rainfall.

Of course, even though the culture "cycles" corresponded with the California sequoia curve in a manner that could not be accounted for by chance, this does not mean that the weather trend in California is always a true indication of world weather, nor does it mean that every part of the world is experiencing the same type of weather trend at exactly the same time. There are always areas in which the weather temporarily runs contrary to the world trend at a particular time. There are also areas that are slowly becoming drier over long periods of time, or gradually wetter. Even in these areas, however, worldwide fluctuations are still discernible.

Ever since Maxwell and Wheeler began making long-range

weather predictions on the basis of predictable successions of gravitational thrusts in Earth's atmosphere, skeptics have stoutly maintained that different parts of the world have separate and different climates. World weather, the doubters have insisted, simply cannot be demonstrated.

The evidence gathered by the Weather Science Foundation and Cyclomatic Engineering demonstrates that world weather is one very complex, but nevertheless single, phenomenon. World weather constitutes a single whole, in which every part is related to every other. In order to understand the rain that is falling in one's backyard, these researchers would argue, it is necessary to know the principles of world weather.

"The first and foremost principle of world weather is *diversity within a unity,*" Raymond Wheeler often stated. The causes of world weather are numerous and complex, but of two general kinds. One set causes unity and uniformity in world weather. These are two quite different things. Weather can be quite different in various places and yet be unified. Another set of causes produces diversity; that is, different kinds of weather at the same time in different parts of the world.

"In other words, as different as temperature or rainfall may be in different parts of the world (diversity), these differences are only specialized parts of a unified whole. All the different parts share in the properties of the whole, just as the brain, stomach, and liver, although different, share in the life of the whole body."

Wheeler's accumulation of materials concerning the history of world weather includes the following data:

- When reports of extreme cold, excessive snow, and freezing and thawing dates are collected from all the known parts of the earth, they cluster in the same decades of history. It is the same with reports of excessive heat, rain, storms, and droughts.
- Heat waves, cold waves, sudden increases or decreases in rainfall—even of two to five years' duration—occur repeatedly over many widely separated areas.
- Large masses of data show that sunspot minima periods tend to be warmer and drier over most of the known areas of the world. Sunspot maxima are colder and drier.

Figure 8 MEASURED TEMPERATURE AND RAINFALL BY CONTINENT

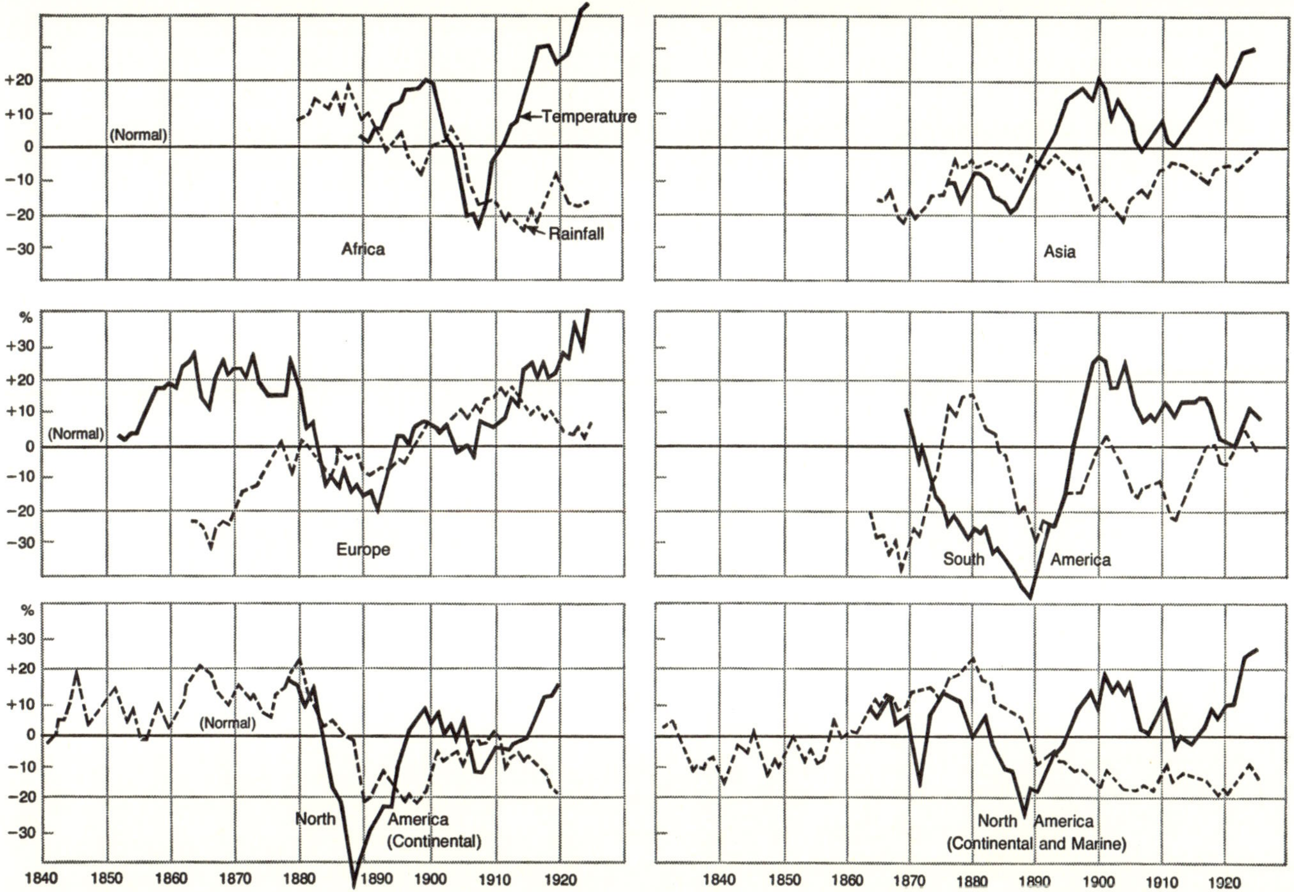

- Masses of data indicate that down through history the warm and cold decades have been so widespread as to be practically universal the world over.

These same masses of data show that the four main climatic phases of the 100-year cycle, disturbed only by secondary leads and lags, occur simultaneously. In cycles as short as 10 years or less, these phases do not always occur simultaneously. One continent or part of a continent will lead or lag behind another.

"The shift from the warm to the cold phase of the 100-year cycle, or from the cold phase to the warm phase, does not occur exactly at the same time over all parts of Earth. Nor will it be equally warm or equally cold each year in countries located at the same distance from the equator. It is the same with rainfall. One part of the world may lead or lag as much as twenty years behind another in reaching the warm side of the average or the cold side of the average, yet discrepancies are the exception rather than the rule. But over a period of years, the great waves of coldness, heat, rain, or dryness will have covered most of the large continental areas, affecting most, if not all, of the longtime averages in much the same way. . . . The majority of the areas will participate sooner or later in the same world trends."

Even when one part of the world is losing or gaining rainfall much faster than another area (i.e., when a local trend varies from the world trend), world trends and cycles of other lengths still show on top of the varying secular trend. Although latitudes near the equator are much warmer and those near the poles are much colder; although actual rainfall varies tremendously in kind and amount in different areas of the world, regardless of the reversal of the seasons in the northern and southern hemispheres; these parts all share in the world weather cycles (uniformity).

The erratic and diversified nature of weather trends during climatic transitions is in itself an illustration of uniformity, since the condition is worldwide.

"The extent of the fluctuations in temperature and rainfall

during the longtime changes in weather trends depends upon the location of the area in question. The warmer countries do not fluctuate to the same extent as do the cooler countries (the tropics least of all); yet they fluctuate approximately at the same time." World weather trends are most uniform during the height of the warm-wet and cold-dry phases of the 100-year cycle, and least uniform during the transitions from the cold to the warm side, or vice versa. During and right after these transitions, the world as a whole receives its maximum rainfall and is, therefore, the most stormy.

"The climate curve is intended to represent—as far as one curve can—the weather trend in the world as a whole at any one time," Wheeler explained. "The curve has no *absolute* significance. The meaning of the curve at any one point is relative to the pattern of the 100-year cycle as a whole. It should be interpreted as follows:

"Here was where the warm-wet phase of the 100-year cycle occurred, and it was broken into by drought years or by drops in temperature at such and such places for approximately the length of time indicated. Here is where the main warm-dry phase occurred according to the evidence, and it was about as stable or unstable as indicated. Here occurred the cold-wet phase, but it turned dry at about such and such a time. Here the apparent hottest years of the 100-year cycle occurred, and here the apparent coldest years; this decade was warmer or colder, or wetter or drier, than the preceding one. Here is where the cycle that averages 100 years in length ended. Here is where a 500-year cycle ended. During this period occurred the longest or warmest of the warm periods of the 500-year cycle.

"The midline across which the temperature and rainfall curves fluctuate represents the composite average condition; that is, normal temperature and rainfall in the majority of the known areas of the Earth."

From data such as these, Wheeler concluded that weather trends are conspicuously synchronized on a worldwide scale, perhaps as much as 80 percent of the time in the long run. Accordingly, a single source of evidence, such as the California sequoias, is a reasonably safe guide about four fifths of the time throughout history

—more when the facts wanted are only the location of the four phases of the 100-year cycle and their major interruptions.

"The correlations between different types of human behavior and the indicated weather trends at the time speak for themselves," Wheeler maintained.

Dr. Raymond Wheeler's extended inquiry showed that all these weather fluctuations paralleled a broad, universal culture pattern that in spite of its vast intricacy, was alternating back and forth as a single unit. Alternations between the mechanistic and organismic patterns which Wheeler had first started speculating about covered all human activity at any time in history and anywhere on Earth. History was found to consist of alternating mechanistic and organismic times or periods. A complete fluctuation in the order of a cycle averaged almost exactly 100 years in length, although it varied considerably from a minimum of about 70 years to a maximum of about 120 years. The majority of the fluctuations were either 90 or 110 years long.

Dr. Wheeler found that during the mechanistic half of the cycle, the emphasis in human activity and thought everywhere was on the *part* of anything rather than on the whole—often at the expense of the whole. In science the emphasis was on the atom; in biology, on the individual organ or the cell; in mathematics, on the infinitesimal, or the individual number. In politics it was on the individual and his rights as opposed to the state, and on laissez-faire rather than on socialism. In psychology it was on the individual "faculty" of the mind—that is, the individual experience or individual response—rather than on the mind or personality as a whole. In art it was on detail rather than on the overall composition.

During mechanistic times art has always been dominated by highly decorative styles like the baroque and rococo. In dress the ornate Elizabethan and Victorian types of costume prevailed. The mechanistic point of view has always generated a respect for minuteness of detail, accuracy of measurement, and great collections and classifications of descriptive facts. The painstaking collecting by the famous pioneer botanist Carolus Linnaeus in the eighteenth century and the work of Charles Darwin in the nineteenth century are typical

Figure 9 THE 100-YEAR CLIMATE CULTURE PATTERN

WARM-WET PHASE

Nation-building
Golden Ages
Best leaders
Revival of learning
Industrial revolutions
Prosperity
New aristocracy
Centralization and unification
High moral tone
Growth of cities
Imperialism
Wars of conquest
Classical, organismic culture

WARM-DRY PHASE

Nation-falling time
Dictators; despotism
Decadence; decline
Persecutions
Massacres
Pogroms
Regimentation
Slavery
Torture Chambers
Socialism
Communism
Absolutism; Fascism

COLD PHASE

Mechanistic culture
Anarchy; secret societies
Individualism
Laissez-faire
Migrations
Travel; exploration
Naturalism:
- Interest in nature
- Revival of the home and family

Psychology:
- Behaviorism
- Associationism
- Sensationism

Revival of religion, evangelism, missions

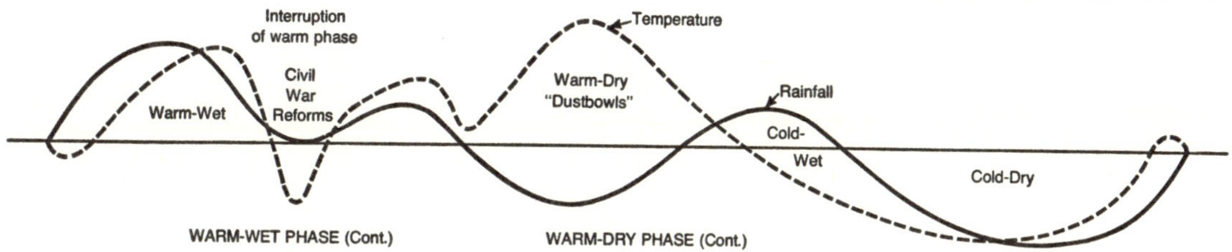

WARM-WET PHASE (Cont.)

Philosophy:
- Rationalism
- Idealism; theism
- Deduction
- Universal laws
- Unity of Science

Geometry
Biology:
- Organism as a whole
- Epigenesis; regulation
- Symbiosis
- Ecology; teleology
- Comparative anatomy
- Physiology
- Integration

Psychology:
- Personality
- Social psychology
- Freudianism
- Topology
- Gestalt theory
- Apperception

Physics:
- Relativity
- Wave mechanics
- Quantum mechanics
- Field forces
- Vector forces

Stereochemistry
Literature:
- Philosophical poetry
- Regimentation
- Free verse
- Social novel
- Pessimism
- Tragedy
- Emphasis on Fate and time

Painting:
- Impressionism
- Subjectivism
- Monochrome
- Abstractions

WARM-DRY PHASE (Cont.)

Statism
Paternalism
Welfare state
Serious depressions
Moral decline
Nudism
Low birth rate
Sex hyperconsciousness
Sexual abnormalities
Fifth columnism
Sovereigns become gods
Surrealism
Fantasy
Caricatures
Murals
Art becomes schizoid
Culture introverted
Fanaticism
Lethargy
Decline in religion
Orientalism in the West
Decline of the family
Dictator-promoted wars

COLD PHASE (Cont.)

Philosophy:
- Empiricism
- Inductionism
- Hedonism
- Materialism
- Atheism
- Deism
- Sophists
- Skeptics
- Agnostics
- Stoics
- Humanists

Biology:
- Preformation
- Cell theory
- Gene theory
- Anatomy
- Surgery
- Pathology
- Taxonomy
- Classification

Literature:
- Lyrics
- Simple rhymes
- Virtues of the individual
- Revolution
- Descriptive novel
- Romanticism

Painting:
- Realism; concreteness
- Naturalism
- Baroque detail
- Rococo; adornment
- Polychrome
- Sentimentalism

Algebra
Lexicography
Dictionaries
Encyclopedias
Geography
Space-mindedness
Wet: prosperity
Dry: severe depressions

examples of mechanistic achievements. On the other hand, the mechanistic pattern has never led to profound *understandings* of nature.

During mechanistic times, causation is sought in the "mechanical" operations of the brain; but during organismic times, causation is sought within the mind itself, as opposed to the brain.

In turn, this emphasis has led to the stressing of dreams, as in the case of Freudian psychology.

The organismic pattern has led to theories of the mind as a whole and to a study of the personality of the whole man and his adjustment to the world in which he lives. Since this world is largely a social one from the standpoint of human problems, the organismic trend has always stressed social behavior and adjustment. Theories of the subconscious have been popular.

During the organismic periods the emphasis everywhere was on the *whole* rather than on the part—often at the expense of the part. These were the periods in the history of science when basic laws were sought pertaining to the operation of the universe in general, such as the law of gravity in the days of Galileo and later the laws of motion of Newton, the gas laws of Boyle, the laws of electricity and thermodynamics in the early nineteenth century, and now the generalized principles of relativity and of quantum and wave mechanics. Basic laws of nature have always been investigated by the organismic, rather than by the mechanistic, pattern of scientific thinking.

In the arts the emphasis on the whole as opposed to the parts has always led to abstract movements—a streamlining of styles —and eventually to some form of impressionism and the caricaturizing of themes. Modernism—the streamlining in architectural design, Impressionism and Cubism in painting and sculpture, and the resorting to fantasies and subjectivism—are examples of the organismic trend in art. Corresponding trends go back to the days of the caveman.

During the organismic periods of history when philosophical and scientific minds were thinking in terms of basic universal interrelations, new nations and governments were being formed. People were becoming more unified, and new governments under strong rulers were becoming imperialistic.

In economic thinking the trend during organismic periods is toward a statism of some kind, either under the rule of an absolute monarch or a dictator or under a welfare, socialistic state. In politics the organismic point of view has invariably generated too great an emphasis upon the state as opposed to the individual. Absolutisms, socialisms, communisms, and eras of dictators have been the outgrowth of misapplications of the organismic pattern.

Developments in the arts and sciences have also come in waves, recurring repeatedly with the cold and warm cycles. Business booms come at the end of a cold cycle and the beginning of the warm-wet cycle which follows.

Human energies increase during cool weather trends. All the waves of rapid advancement throughout history have come in the cool cycles, or as a result of the cool cycles.

Human energies decrease during the warm cycles and bring decadence in all human affairs.

Increased production resulting from improving growing conditions occurs during the warm-wet cycle. Birthrates, however, decrease in warm times; and as the warm period advances, human energies are depleted even more. The effect of the lowered energy gradually produces an economic depression.

Still, Wheeler's cycles would be of only modest use if they only correlated with changes of "mood" or taste. What is striking is that they can also pinpoint specific *events*.

Throughout history, dictators, tyrannies, and the major wars between strong nations have always come during each and every warm cycle. Throughout history, civil wars and swings to democratic forms of government have repeatedly come on each cold cycle.

The chart in Figure 10, depicts the major events of history, graphically, in the form of a curve. The positions of the peaks and troughs of the curve were determined by the distribution of the wars of world history that have occurred since 600 B.C. Every dated battle (unduplicated) mentioned in some 250 volumes of history was used. This means that practically all of the known, dated battles fought

Figure 10 DR. RAYMOND WHEELER'S "DROUGHT CLOCK"

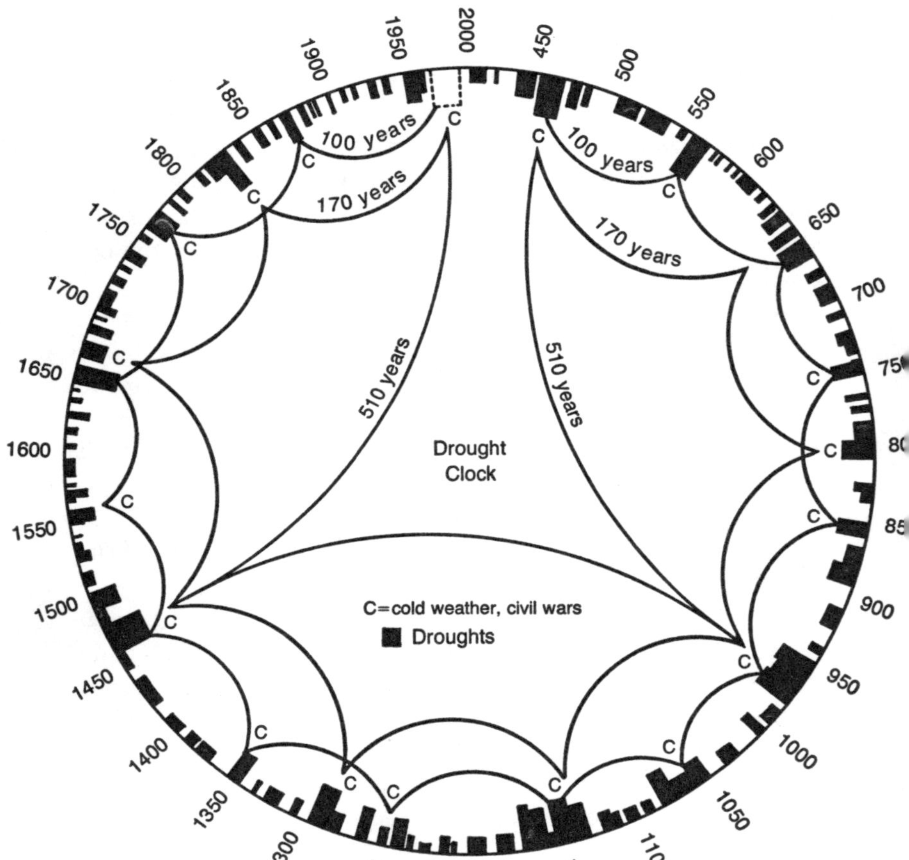

According to his extensive research, Earth's climate shifts from warmer to colder and back again in rhythmic cycles. Nations are built on the shifts from cold to warm, and fall on the transition from warm to cold. Most international wars are waged during warm periods, civil wars during cold times. Each climatic phase begins wet and ends dry.

Cold droughts and centers of civil war epochs usually coincide, with a major cold drought and a period of civil strife occurring about every 510 years. Less severe cold-dry times occur every 170 years. Dr. Wheeler's work indicates that we have just entered a major cold drought era—and a time of great civil unrest and social revolution.

anywhere in the world since 600 B.C. were included. After these conflicts were divided into international and civil, Raymond Wheeler found that the two kinds of wars alternated with one another; that is, there would be a time in history when international wars prevailed, than another when civil wars prevailed. If a war happened to be a mixture of the two, half of it was recorded as international, half as civil.

Entries or tallies were made of the number of international battles. These tallies were arranged above the midline, and the heights of these tallies were made proportional to the severity of the battles. Under the same plan, tallies for the civil war battles were arranged below the midline. In this way the positions of the peaks and the valleys were determined exactly by the dates of the battles fought in the wars.

In fact, Wheeler discovered that the dates of battles alone would have been sufficient to locate warm and cold periods of history with striking accuracy.

After the curve had been drawn, it was found that the peaks and the troughs based on the wars of history coincided in an astonishing manner with the peaks and troughs of the composite sequoia tree-ring curve, based on trees growing in both dry and wet situations on the western slopes of the southern Sierra Nevada in California. In this region the seasons repeat themselves in such a regular manner that the sequoias almost always lay down one ring a year. These rings can be counted, and the age of the tree determined.

There are long periods of time when the average rings are relatively wide, and periods when they are relatively narrow. These periods alternate. The periods of wider rings are represented by peaks in the curve, and the periods of narrow rings by the troughs. These peaks and troughs coincide in a close manner with the peaks and troughs of the wars of history when divided into international, above the midline, and civil, below.

It has been shown that the longtime peaks of the sequoia curve occur during warm times and the troughs during cold times. The growing seasons are longer during warm years, hence the rings are wider. But the rings are also determined by the amount of rainfall. The rings are wide during wet years and narrow during dry years. They are the widest during periods in which there is a generous amount of warmth and moisture, narrowest when it is both cold and dry. The dating of the peaks and troughs of the economic-culture-climate curve was therefore determined jointly by the distribution of wars and by the dating of positions of the major peaks and troughs of the tree-ring curve.

The curve in the chart has been adjusted to represent three related facts at the same time—the distribution of wars, fluctuations in tree growth, and temperature.

In regard to temperature, the story as told by the trees in relation to known fluctuations in weather in recent centuries was applied to the earlier centuries and thoroughly checked against thousands of individual items found in various documents of the past, that is, actual weather events such as heat waves, cold waves, storms, blizzards; crop failures, famines, locust plagues, harvesting dates; dates when ice broke up or formed on lakes and rivers; lake and river levels; and so on.

It turned out that when international battles were plotted above the midline and civil war battles below, the resulting curve provided an excellent representation of the rise and fall of empires, nations, and governments throughout history. On the peak (or warm) side, where international wars prevail, empires are strong and unified and prosperous enough to wage these wars. On the trough (or cold) side, where, along with anarchy, civil wars prevail, governments have collapsed. Factions among the people of different political or religious beliefs tear the country apart. (Out of this chaos, however, emerges a new national feeling in a new generation of people under fresh leadership. The movement accelerates as the cold period comes to an end and reaches a climax just as it is turning warm.)

When compared with curves of measured temperature and rainfall, their distribution showed that international wars were typically warm-phase events, while civil wars occurred most often during the cold phases. The curves showing the distribution of these battles over the centuries would have located the warm and cold periods of history for the world as a whole.

The famous civil wars of history have *always* occurred during the long cold periods: for example, the civil wars between the aristocratic and democratic parties in the Greek city-states, circa 400 B.C.; the civil wars of the Gracchi in Rome in the early 100s B.C.; the civil wars between Caesar and Pompey, then between Anthony and Octavius in the 40s and 30s B.C.; the civil wars all over the world in the second half of the 400s A.D., typified by the official fall of Rome (and of all the other ancient empires as well). In more modern times there have been the civil wars between the English houses of Lancaster and York in England (the Wars of the Roses), which occurred during the cold parts of the fifteenth century and the fierce religious wars in Holland and France in the 1500s, which came during cold decades, as did the Reformation. The American and French revolutions broke out at the end of a long cold period. The South American wars of independence fought under the leadership of Bolívar occurred during cold times, as did the American Civil War. Because we have just ended a warm period and are well into a cold period, we are now beginning a period of potential civil wars. No wonder civil war erupted in Vietnam. The difference between "hot" international wars and "cold" civil wars makes better sense when we recall the natural progression from organismic (hot) politics to mechanistic (cold) politics and back again. Whenever an organismic political whole overemphasizes the state at the expense of the individual there invariably develops a wave of socialistic and communistic theory, along with the emergence of tyrants, dictators, despots, police systems, reactionism, and a contraction of all freedoms that had evolved up to that time in history.

With the growth of strong governments, there occurred an outburst of international wars of strong against strong. This was when the culture pattern was shifting as a whole from the piecemeal emphasis to an emphasis on wholes. The outbreak of international "nation-building wars" are invariable symptoms of the revival of nation building following a period of political (mechanistic) stagnation. Periods of nation building, coinciding with the reappearance

Figure 11 CLIMATE AND HUMAN BEHAVIOR

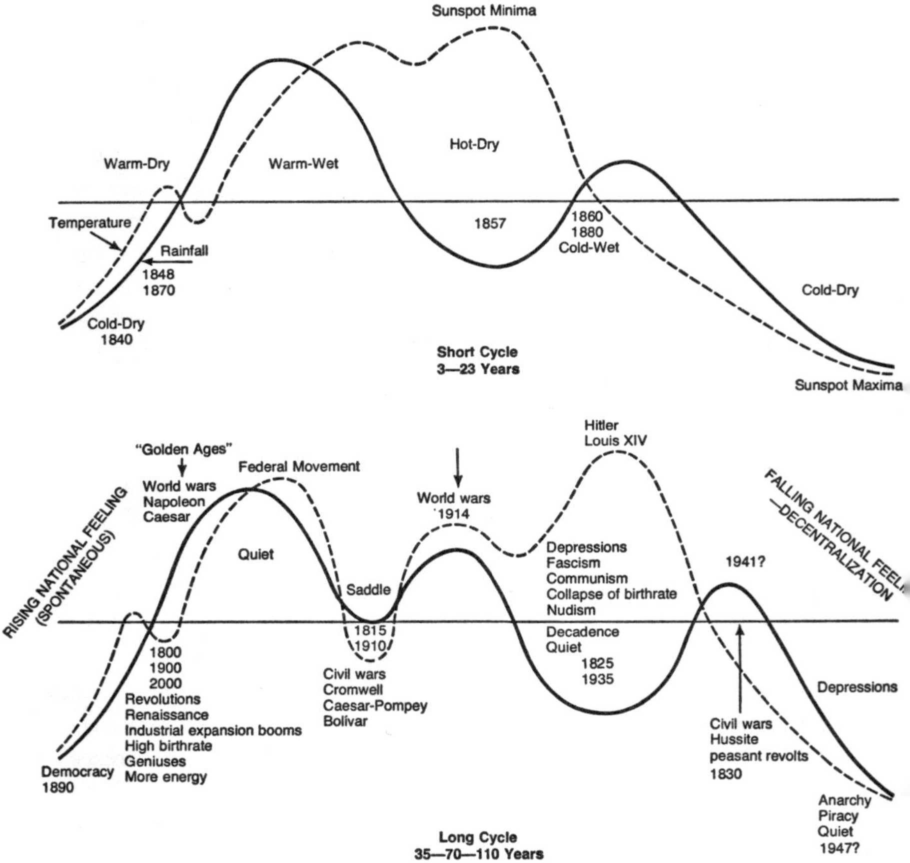

of the organismic culture pattern as a whole, invariably end in periods of social, political, and moral decline, and a period of universal despotism.

At this point in the cycle, nations are on the way to collapse, or at least governments are about to break down. As part of the breakdown process, there occurs another wave of international wars which can be called "nation-falling wars," because they are symptomatic of political decline or degeneracy, or at least of political extremism. These wars occur on the reverse transition; that is, as the organismic shifts back to the mechanistic pattern. They give way to civil strife in the form of revolts, rebellions, and other extreme movements toward the Left.

People all over the world, whose suffering under an autocracy of one kind or another has reached a climax, now possess the initiative and determination to free themselves and to revive and extend the democratic institutions they had known previously, but which had all but been lost during the recent trend toward statism. Several decades are then taken up with the process of emancipation. During the process of nation building, a new aristocracy, or minority party of some kind, had seized power. This privileged class is now overthrown, and participation in the government is extended to a larger percentage of society. A new generation sets up a new, more democratic government—only to initiate another trend in the direction of absolutism and tyranny as the culture pattern shifts once more from the mechanistic back to the organismic emphasis.

Thus it is that the civil wars of history have been associated with the mechanistic pattern, and international wars with the organismic.

5

Known Effects of Climate on Human Vitality

It seems obvious that there are such patterns in history. But if they have to be correlated with anything, why weather, specifically? Is man's behavior really linked to humidity and temperature?

"Look," Cejka said, his tone serious, "there is the accepted fact that plants and animals, all through the evolution of life, have been in part the direct products of their environment. That they have always been integrated with that environment in their form and function. Why should this not be true of man?

"Why, Brad, the simple truth is that man has not escaped! If environment is a strong enough factor to condition the structural features of a plant or an animal, how much more must it determine directly and indirectly the behavior of the nervous system, the habits, temperament, attitudes, tolerances and intolerances, intelligence, and even the thoughts and the aesthetic life of human beings?"

"The climates of the different areas of Earth produce different kinds of plants and animals," Wheeler observed. "No one disputes this fact. Their physical characteristics, their habits, and their behavior are determined in large measure by the climate in which they live. When the climate changes, the plants and animals must adjust or die. Dramatic changes in climate are what has caused so many species to disappear from Earth in the past.

"Man is as much a part of the life of Earth as are plants, crops, and animals. One would not expect man to be immune to the conditions that affect and influence plants and animals, and he most definitely is not."

John Cejka paused, tamped the tobacco in his pipe, and reached for a fresh match. "You know, Brad, this interest in human ecology is no new thing. J. W. Bews wrote a book entitled *Human Ecology* back in 1935. Bews said there was a pattern that influences the whole life of the individual—from adolescence to marriage to adult life to burial behavior. He said that the type of society which can exist economically and politically is influenced by climate.

"Whether a people become herdsmen, pastoral nomads, fishermen, craftsmen, artisans, traders, merchants, or so forth is directly or indirectly the product of the geography and climate of a given region. Bews stated that an *unchanging* climatic environment could do nothing but produce stagnation in man's habits and culture, while on the other hand, *changing* climate conditions will lead to evolutionary progress—or if too great, to destruction."

"Of course, John," I agreed. "It's only common sense to assume that people living near the ocean aren't likely to become pastoral nomads."

"But it isn't that simple," John said, smiling. "Why is it that some people who live near the sea raise themselves no higher than the level of primitive fisherfolk, while others build marble temples and become seafaring merchants?

"Over a century ago," Cejka went on, "H. T. Buckle wrote that the energy and regularity with which labor is applied in the process of building a civilization determines the accumulation of wealth, *but it is the climate that determines the energy and the regularity of the laborers!*

"In a temperate climate the worker is invigorated; in a hot climate he becomes very lackadaisical; in an excessively cold climate he is prone to desultory habits. How cold does it get in Decorah in January?"

"About twenty, thirty below zero."

"And when the snow is a foot deep, wouldn't you rather be home in front of a crackling fire?"

"Yes," I admitted.

"Exactly. Long seasons of cold break the chain of industry."

Temperatures do have a direct influence upon the physiology of the body. The human body is like an engine that must convert one form of energy to another. In doing so, like any other engine, it creates heat. The body must get rid of this heat. Ordinarily it does so by radiation. However, when outside temperatures are high, radiation is often too slow. Unless the entire body is slowed down and generates less heat, fever and death will result.

When temperatures are high, the body tissues swell and contain more water. The blood vessels open up to help the escape of heat; the blood turns acidic; blood pressure falls; breathing is more shallow; the body receives less oxygen and loses weight. The nerve reactions are speeded up. Thrombosis is more frequent, due to sluggish circulation.

If the body stays hot enough, it has to remain at a low level of activity. There is no reserve energy for extra accomplishments or hard work. Growth slows down; the body is less resistant to disease; progress is slow; fertility low. Action must be less vigorous. More time must be spent resting, hence the siesta in warm climates.

Dr. William F. Petersen, a pioneer investigator of the influences of weather on the etiology of disease, was interested primarily in the medical side of climatology. According to his observations, there is a tendency for introverted personalities to prefer warm weather, and extroverted personalities to prefer cold weather. Introverts, whose minds turn inward and stress thinking and imagination, sympathy, and the importance of *security*, are less happy and stable during cold times and get tense and unstable during cold waves. Introverts like warm weather, and they stabilize on rising temperatures (until it gets too hot). Extroverts, whose minds turn outward and stress action, and competition, practicality, and freedom, are less happy and stable during warm times and get tense and unstable during heat waves. They like cold weather and are more stable on falling temperatures (until it gets too cold). Weather that makes the strong stronger makes the weak weaker.

In cool weather or in cool climates, blood pressure is higher; hearts beat faster; glands are more active; metabolism in general and growth are faster. The blood vessels constrict to save heat; fat develops in the interest of conserving body heat against cold outside temperatures; and animals and human beings alike are more alert, aggressive, stable, tolerant, and intelligent. The same conditions that produce action produce higher I.Q.s, more competitive spirit, faster technological progress, and greater tolerance.

On cool days most people feel more alert and alive; they have more push and drive; they have more optimism, greater endurance, greater stability, and feel more competitive. Salesmen who make their calls after a storm when it is cool and the barometer is rising will find the buyer more receptive and more tolerant. On cool and stormy days more people get to work on time, and there are not as many absences.

But it must not get too cold. Cool weather and cool climate have their disadvantages. Your ulcers are likely to act up or perhaps your appendix or even your gallbladder. When it gets too cold, man becomes retarded and less stable.

Dr. Wheeler went on to illustrate that experiments on animals show that there is an optimum temperature for the development of size, stability of behavior, fertility, and intelligence. There are optimum temperatures for the most economical conversion of energy from food into meat, as in hogs and other animals.

According to the research data of Dr. Ellsworth Huntington, there is also a best temperature for human intelligence and vitality. "To point out only one study made of the efficiency of factory workers in Denmark, Japan, Pennsylvania, New York, Maryland, the Carolinas, Georgia, and Florida," John Cejka told me, "it was found that there was a pronounced peak of physical activity in the spring and fall, with valleys in midwinter and midsummer. Corresponding variations were found in mental output.

"The scholastic grades in mathematics of West Point Cadets, the percentage of individuals passing civil service examinations, the scores of students taking college entrance exams—all show spring and fall highs, midsummer and midwinter lows. Huntington looked at these results and believed that they pointed to temperature optima for physical and mental activity. He set them at 38° F. outside temperature for mental work, 68° to 70° F. for physical activity."

For the "best" effects there must be a moderately large difference between summer and winter temperatures and between night and day temperatures. An active, competitive civilization cannot thrive where it is warm the year around, where the nights are about as warm as the days, or where rainfall is not properly distributed throughout the year.

"Then only a relatively small land area of Earth has this climate best-suited for human progress and technological advancement," I said.

Cejka nodded vigorously. "Parts of the earth where these conditions are fulfilled to the greatest extent are the parts occupied by the highest levels of civilization, as measured in terms of scientific and technological progress and in standards of living. Today the optimum temperature belt for maximum human vigor runs through Western Europe—east to the Ural Mountains and the Black Sea—and Great Britain; across the Atlantic to southern Canada; the northern United States from New England through the Great Lakes region; and the southern United States during cold periods; the Northwest; then to Japan. With some qualifications, the 'good' area extends eastward across the Balkans and southern Russia as far as the Black Sea, and to a lesser extent, a portion of central China, parts of Korea and Manchuria; northern Argentina and central Chile; the southeastern part and southern tip of Africa; the southeastern corner of Australia and New Zealand. The best climate for man ranges from

47° to 52° mean annual temperature, but 40° to 60° is good. Very much beyond this range and it becomes either too cold or too hot for human vitality, intelligence, initiative, and progress. After the temperate zone expanded, the southern half became too warm —some too dry, other parts too wet—for maximum human vigor. This is now the condition across the Mediterranean, Asia Minor, India, southern China, and Central America."

"The very places where the high civilizations once existed," I reminded him.

Cejka smiled agreement. "Yes, and let's go on. North Africa and Turkey have declined because it has been too warm and dry in those regions for several centuries. On the other hand, Great Britain became a great empire because the climate of the British Isles is about as good as any in the world for a steady maintenance of aggressiveness and vitality. The only areas capable of supporting large populations are in North America and Western Europe.

"Germany became a powerful nation because of a stimulating climate, while Poland and Russia are most of the time too cold. China has never been able to unify because of her variety of climates."

The sheet of paper between us on the desk was marked with doodles, numbers, arrows. John Cejka thinks and speaks better when his hands are occupied with providing supplementary visual aids to the points he wants to emphasize.

"I suppose one has to be very careful how he chooses his words when he speaks of the people of one climate being more vigorous than those of another," I observed. "You could find yourself in the sticky areas of alleged racial superiority or nationalistic stereotypes."

"But this has nothing to do with peoples, with races!" Cejka said emphatically. "It all has to do with climate, temperature, and weather-energy cycles! Now we know—and this has nothing to do with race—that people living in extremely warm climates are not as progressive or healthy as those people living in cooler climates. The cultural differences here in the States are climatically conditioned. The South is too warm and dry for maximum vigor. In other words, our good friends and neighbors from the Deep South are not lazy.

They have to move and achieve more slowly or become ill and die. On the other hand, provided they remain quiet enough, they will ordinarily live longer because they are burning their candles more slowly. Even their heartbeats are slower."

Dr. Wheeler explained that the human organism must be very delicately adjusted to the prevailing temperatures of a given region. The delicacy of this adjustment is all the more remarkable in light of the fact that people living in the same country, under annual temperatures, say, of 46° F., as in Minnesota, carry on in the same political union in relative harmony with people living under annual temperatures of 66° F., as in Louisiana—a 20° difference. In both groups, attitudes of mind, tolerances, and intolerances *will change in the same direction, but perhaps in different amounts, with not over a 3° change in annual temperature, in each region.*

In any known region of the earth, the most extreme difference in measured annual temperature between the coldest half decade and warmest half decade, up to 1930, does not exceed 6°. Most of the differences are considerably less than that. But a shift in annual temperatures of that amount may be the equivalent of a considerable shift in latitude. A difference of 3°, in any region near the middle of the temperate zone, is going to change the relative length of summer and winter considerably. During cold years, which are often dry, the summers may be hot, although shorter, so that the differential applies mostly to the winter months. Hence the effect on human attitudes might be all the greater.

"To be sure, there was a time when cultural differences between the North and the South in the United States, obviously conditioned by climate, were the cause of a bloody civil war, for slave civilizations have never flourished well in cold regions, or in warmer regions during long cold periods," Dr. Wheeler admits.

"The difference of 3° and less, almost shockingly small, was probably much greater during earlier centuries of history when, perhaps, the warm phases were warmer and the cold phases colder than now. Nevertheless, a difference in mean annual temperature of no greater than 1-1/2° F., when prevailing consistently for no longer than half a decade, is sufficient, anywhere on Earth, *to start* changes in the human behavior pattern in one direction or the other."

Dr. Wheeler's research data showed that it made no difference whether human beings lived in colder or warmer countries, in the Occident or the Orient. Their political behavior shifted in the same direction with the same shift in world climate. This was true regardless of level of culture, or the degree to which the political unit in question was more or less democratic or totalitarian generally than another political unit.

"It seems likely that the difference in temperature alone will not account for the differences in human mental attitudes," Dr. Wheeler notes, "but that other climatic events which accompany rising and falling annual temperatures are important. . . . Storminess was associated, over the world as a whole, with climatic change, and when temperature had stabilized at an extreme level on either side of the axis, it was dry—that is, storms were much less frequent.

"Under storms, of course, may be included ordinary rains. Thus, changeableness of the weather and, with longer periods of time, changeableness of climate go with temperatures that approach the longtime mean. The more invigorating and changeable climate (as to both temperature and storminess) occurs while annual temperatures prevail near average for a given region—that is, are near the average temperatures to which the human race is adjusted, at a given time and place. In the active areas of Earth, these temperatures are near what have been termed the 'temperature optimum for human vitality.' "

"You know," John said, laughing, his enthusiasm revving even higher, "Maxwell ran one study in which he took some three thousand names from *Who's Who*. He took the birthdays of these outstanding people, charted back three hundred days to cover the average point of conception, then matched these dates against what his weather-energy charts said were superior times for conception."

"That sounds a bit like a kind of super-astrology," I ventured.

"Even so," Cejka went on, "Maxwell's research revealed that there were eighteen times more birthdays of these notable people in periods designated as good than there were in the poor periods. Maxwell went further. He charted the dates that would produce the best plants, the largest litters, when males or females were most likely to be conceived. . . ."

As a part of his research on the influence of weather trends and cycles upon human affairs, Dr. Raymond Wheeler was once asked to comment on the matter of birthrates. His main points were as follows:

- The conception rate varies with temperature. Sterility in both sexes is produced by excessive temperatures. The reproductive cells, especially those of the male, have less vitality at such times and develop defects.
- Certain hereditary factors, associated in the past with climatic factors, are also in the picture.
- The birthrate also fluctuates with climatic change. Spontaneous high birthrates, both in man and in animals, occur under certain definite climatic conditions and under no other conditions. The highest birthrates occur when it is cool or cold and wet, but not too cold. The best times for high birthrates are during climatic transitions from the warm to the cold periods and from the cold to the warm periods in the 100-year climatic cycle. These are times of highest vigor, energy, and best health of the reproductive cells. They are also times of greatest interest in children and family life. They are times of large families.
- The same high energy level that produces an increased birthrate produces optimism and aggressiveness in the business world. As a result of this drive and confidence, competition increases, production increases, and prices rise. This is boom time.
- The lowest birthrates occur when it is excessively hot and dry (with little or no relief during the night) and when it is warm and moist (humid) for long periods of time. Periods of great depressions and low birthrate (low birthrate not basically caused by the depression) coincide because of lowered vitality of the human races at such times. Animals suffer lowered vitality at the same time.
- Those conceived in May or June lead more stable lives and accomplish more than those who were conceived in July—they also live, on the average, four months longer. Those conceived in summer, in hot desert areas, died, on the average, ten years sooner than those conceived in the same area in early spring.

"And all these findings were based on the weather-energy cycles?" I asked.

"Maxwell found that gravity had the effect of increasing vitality in all living organisms," Cejka said. "The weather turned out to be the *x* factor! The old-time farmer may not have had more than folklore to guide him when he planted according to the cyclic phases of the moon, but now we can see that there was a very sound scientific reason behind such programming. The good days to plant can be calculated in advance."

The eminent American scientist and historian John W. Draper said over one hundred years ago: "Where there are many climates, there will be many forms of men. . . . For every climate and, indeed, for every geographic locality, there is an answering type of humanity."

Different climates of the world produce different kinds of people, just as they produce different kinds of plants and animals. The size, the color, the intelligence, the emotional stability, the tolerances and intolerances, the philosophies, and the moral caliber of peoples are all basically determined by the climate or the prolonged weather trends to which they are subjected.

"Physicians, historians, philosophers, geographers, and statesmen in almost every century of history, since the time of ancient Greece, have recorded their opinion that man is profoundly affected by the climatic characteristics of his environment," Raymond Wheeler once wrote. "The interesting part is that these numerous observers have been in almost complete agreement regarding the main effects of climate on human nature.

"For example, they have agreed that people living in cool, temperate climates (not the frigid cold of Arctic or near-Arctic regions) are more vigorous, more aggressive and progressive, alert and persistent, stronger physically, larger, braver in battle, healthier, and are less prone to sexual indulgence, although they have more children.

"In warm climates mankind was found to be more timid,

smaller, physically weaker, and less courageous, but more inclined to physical pleasures, including sexual indulgence; that they were more effeminate, lazier, and less aggressive as well as less progressive. Scientists of the ancient and medieval worlds over a period of almost two thousand years expressed much the same opinion.

"No one doubts the correctness of this picture, in general, of the differences between cool (not too cold) climate and warm climate races and nations," Wheeler contends. "It is easier to reason with cool-climate peoples; they are more stable in their behavior and much more democratic, other things being equal. Every historian knows that slavery has never flourished in cool climates, but only in warm climates (unless in cool climates during excessive, but only temporary, warm periods). It is common knowledge that the brutalities inflicted upon underprivileged minorities; the bloodthirsty mutilations and tortures carried out among the members of royal families in the struggle for personal power; the bitter jealousies; the settling of differences by duels rather than by arbitration and compromise; the stabbings and other forms of murder, often on slight provocations; and cruel sex crimes of sundry kinds are all far more typical of warm countries than cool countries, but will reach a climax in cool countries during warm periods.

"Since ancient times it has been observed that warm-climate peoples were more bloodthirsty and given over to murdering or slaughtering one another than were cold-climate peoples," Raymond Wheeler states. "We have noted that there is a strong tendency for state-promoted persecutions, pogroms, and massacres to occur during the warm-dry phases of the 100-year cycle. A graphic example is the horrible treatment by German officials of Jews and political offenders in Poland and Germany preceding and during World War II.

"Tabulate the palace murders of history. A large percentage of them will be seen to have occurred in the warmer, Oriental countries. This is especially true of mutilations and torturings, where one member of the royal family caused another to be blinded or otherwise mutilated. It is also a well-known fact that more irregular sexual practices occur in warm than in cool countries. It would appear that these irregularities increase among the inhabitants of the cooler countries during the warm phases of the 100-year cycle and decrease during the cold phases. In other words, there are grounds for believing that sexual activity is more orderly during cold periods and among cool-climate peoples.

"Warm climates or warm weather trends affect the sex life of people in two ways:

"First, heat renders organisms more sensitive and delicate. Warm-climate peoples are less stable and more emotionally intense.

"Second, they are more passionate, yet, on the whole, less fertile. They are more sex-conscious. On the other hand, sex is taken more for granted in cool-climate peoples.

"There are different kinds of people living in different climates, with different habits, different mental and physical characteristics, and different temperaments and philosophies of life. There are marked differences between people living in the warmer and the cooler countries. Indeed, man's level of intelligence, his fertility, and his energy level depend upon, and fluctuate directly with, the changes in the weather trends."

"Isn't Wheeler also saying that climate shapes civilizations in the same manner, regardless of which continent supports the culture?" I asked.

Cejka nodded. "Whenever the climate of two cultures is similar, regardless of how distant they may be from one another, they enact scenes of analogous historical developments, even to economic attitudes and political behavior in general."

"I've heard that Los Angeles and Jerusalem have the same climate," I put in.

"A theorist by the name of Demolins once said that if the history of humanity were to begin over again with the same biological factors and with the surfaces of Earth unchanged, history would be repeated in its larger outlines," Cejka went on. "The same trade routes, geographically determined, would determine the social and economic contacts of one race with another. These contacts would, in turn, create the same social and racial types. From this point of view, geography determines the character and evolution of human societies.

"Effective progress depends upon man's energy. Climate reacts upon man to stimulate his energy and to develop his powers."

"But," I wondered, "can't man somehow rise to the challenge presented by his climatic environment?"

"Absolutely!" Cejka agreed. "Civilizations conditioned by human effort have always outstripped those which have depended mainly upon nature alone and have not summoned the latent powers of man to meet the challenges which nature presents."

I nodded. "It has been said that the British and the New Englanders make good horse traders because the weather they are subjected to is mean enough to make them unhappy, uncomfortable, and aggressive. Competition meets the challenge."

Amusingly enough, Maxwell and Wheeler predicted in 1935, that a very marked interest in world weather, weather cycles, and their effects on people and life in general would emerge when Earth began to get colder. Because of the effect of cycles on all human affairs, they knew that recurrent interest in weather has occurred all down through history. Weather and climate consciousness has always been associated with an outburst of travel, exploration, and field studies.

Wheeler found that scholars in the seventeenth century, during long warm periods, had relatively *little* to say about the effect of climate on human beings. When it is warm, people lose interest in their environment and become more interested in themselves. They become more subjective and introspective, as opposed to being objective and environmentally minded. During warm times, people are *time*-minded and social-minded, as Wheeler put it. "They have, then, a subjective point of view."

"Now, when people are social-minded, they are interested more in *themselves* and in *security*; they are interested in the *welfare state*. This has been so, regularly, down through history during warm times, so that times of socialism, communism, fascism, and like types of government under despots are predictable." During the warm periods of 1790-1830 and 1900-1940, weather cycles were neglected and frequently denied.

"It is significant that the times in history when the most interest has been shown in this problem have been the cold parts of the 100-year cycle. This is because people are environment-minded during cold times (they have an objective point of view). Great naturalists such as Charles Darwin have all been cold-phase, *space*-minded men and women. During the cold periods of the eighteenth and nineteenth centuries, weather facts were recognized and studied. When people are environment-minded, they are interested in climate and weather, and their effects; they are interested in *space*—travel, exploration, migration."

And rainfall and temperature records from various parts of the world show that Earth as a whole has been getting colder and drier for a good many years.

PART II

UNRAVELING THE SKEIN OF THE PAST

6

The Warm-Wet Phase of the 100-Year Cycle

Raymond Wheeler and his research associates found that throughout history the weather has had so much bearing on the lives of mankind that it has strongly influenced people's attitudes, thoughts, moods, temperament, activities, and behavior. The weather directly controls all human activities, causing men and women to run in cycles in direct conformity to the weather cycles.

So many aspects of human behavior are covered in Wheeler's massive digest of world history that he developed a guide which explained particular patterns of behavior in different phases of the longtime weather cycle.

On Figure 11, the tops of the peaks and the bottoms of the troughs represent drought times. On the warm side the droughts continue as the curve falls toward the midline. The curve is dropping here because temperatures are falling, but also because the longtime average width of the tree rings is lessening on account of the droughts. The downward slant of the curve represents both facts—falling temperatures and the slowing of tree growth. The periods of droughts on the warm side are shown by shading.

The heights of the peaks are proportional to the temperature as indicated by the trees and by direct evidence from documents. The depths of the troughs are proportional to the severity of the cold period, as indicated by similar evidence.

Wheeler did not interpret the cycles as indicative of *complete* environmental determinism, but man evidently obtains energy (physical and mental) and motivation of an imperialistic, totalitarian, and absolutistic character on one hand, and of a nonimperialistic, individualistic, and democratic character on the other—in part at least through the way in which he is integrated with his natural environment. The one pattern flows over to the other, inevitably, as climate shifts back and forth between warm and cold periods.

Types of Behavior during the Transition from Cold to Warm

**A period of high human energy level.* Wheeler's intensive study makes it appear obvious that something about opposite extremes of temperature produces opposite extremes in collective human behavior. The association of these extremes of temperature with dryness or lack of action in the environment has something to do with human decadence. The shift from a long cold period to a warm one—or, better, the sequence of these two conditions—has something to do with an increase or release of human energy that produces "better" modes of behavior. It hardly seems coincidental that hell is supposed to be a "hot" place, and heaven balmy and springlike.

"With increased vigor as a base (whatever the physiological causes may be), optimum conditions for an abundance of available energy for work occur during the period of climatic normality and on the upward crossing, or transition, from cold to warm," Dr. Wheeler declared. "This is the 'springtime' of the climatic cycle, while the preceding cold period was the 'wintertime.' On the upswing, more than in any other place on the cycle, the human race possesses energy, above that necessary for a maintenance of his physiological engine."

**Maximum of human vitality as revealed in aggressiveness, longevity, high birthrate, and low incidence of illness.* "Here, mental and physical energy alike are at a maximum: hence the appearance of both good leadership and good followership; economic and political aggressiveness and enthusiasm; ability to exercise more self-control and make better judgments; predominance of constructive measures; and the absence of decadent modes of behavior. With all of these are associated a greater incidence of genius, a generally higher birthrate, a more stable behavior, and a higher moral tone of society."

**A temporary lengthening of the sunspot cycle and a reduction in the number and size of spots for the duration of the warm phase; the beginning of a rainfall maximum with a long period of good crops; storm and flood maximum.* "Moreover, physical conditions are then the most favorable for economic prosperity and for the growth of stationary societies, dominated by city life, for rainfall is ample and crops are good."

**A great majority of the best leaders in history: better than 90 percent of the sovereigns who have been given the title "the Great".* The names on the chart are primarily those of the most famous sovereigns of history, known for their constructive and brilliant rules.

Ninety-five percent of the rulers who have come down through history with the title "the Great" have ruled at this time. Eighty percent of the rulers positioned at this place on the curve have been judged good by historians. These are the leaders of the revivals of civilization, the birth or rebirth of empires, nations, and governments.

Nearly all of the names on the chart are located on the upward slant of the curve, between the midline and the rising side of the peak. This is nation-building time, a time of prosperity, a time of the best governments, which come at the midline on the upward

Figure 12

THE 100-YEAR CLIMATIC AND CULTURE CYCLES OF HISTORY

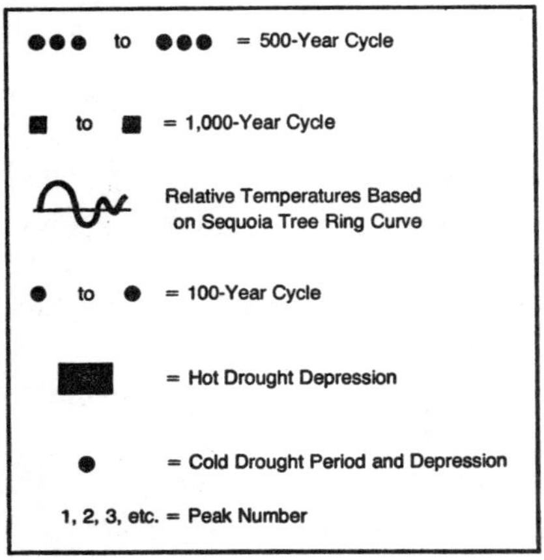

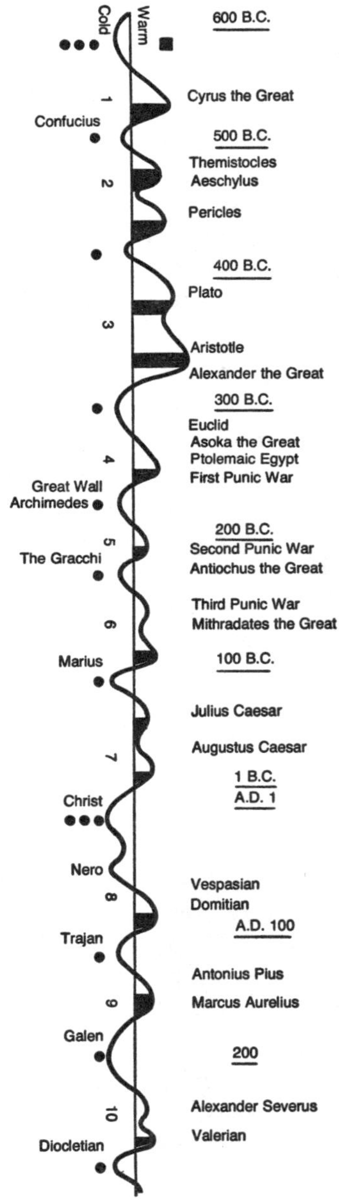

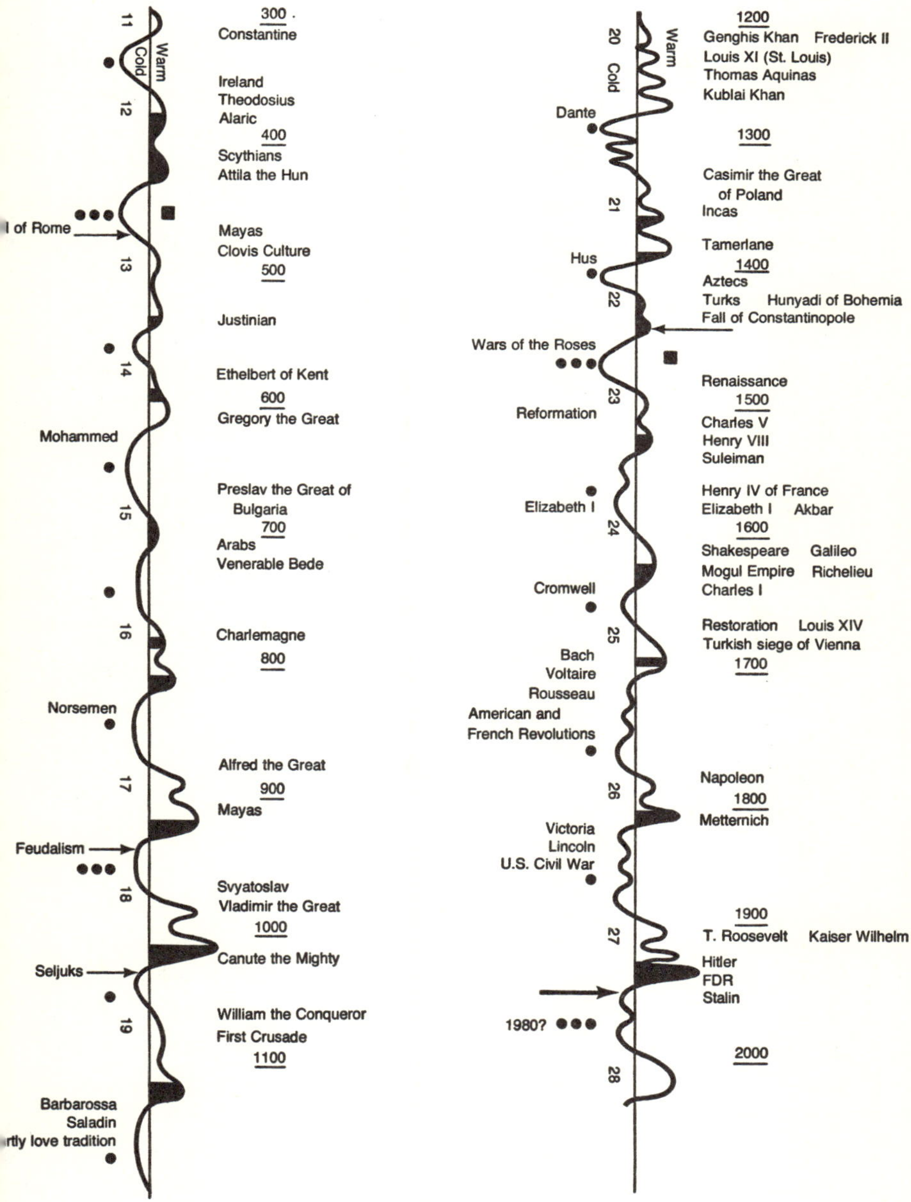

Warm
Cold
11
12
13
14
15
16
17
18
19
300
Constantine
Ireland
Theodosius
Alaric
400
Scythians
Attila the Hun
l of Rome
Mayas
Clovis Culture
500
Justinian
Ethelbert of Kent
600
Gregory the Great
Mohammed
Preslav the Great of Bulgaria
700
Arabs
Venerable Bede
Charlemagne
800
Norsemen
Alfred the Great
900
Mayas
Feudalism
Svyatoslav
Vladimir the Great
1000
Canute the Mighty
Seljuks
William the Conqueror
First Crusade
1100
Barbarossa
Saladin
rtly love tradition
20
21
22
23
24
25
26
27
28
1200
Genghis Khan
Frederick II
Louis XI (St. Louis)
Thomas Aquinas
Kublai Khan
Dante
1300
Casimir the Great of Poland
Incas
Tamerlane
1400
Hus
Aztecs
Turks
Hunyadi of Bohemia
Fall of Constantinopole
Wars of the Roses
Renaissance
1500
Reformation
Charles V
Henry VIII
Suleiman
Henry IV of France
Elizabeth I
Akbar
1600
Shakespeare
Galileo
Mogul Empire
Richelieu
Charles I
Cromwell
Restoration
Louis XIV
Turkish siege of Vienna
1700
Bach
Voltaire
Rousseau
American and French Revolutions
Napoleon
1800
Metternich
Victoria
Lincoln
U.S. Civil War
1900
T. Roosevelt
Kaiser Wilhelm
Hitler
FDR
Stalin
1980?
2000

swings of the curve and last through the first part of the warm periods, while it is still raining.

**End of a civil war period and the beginning of an international war period associated with rising national spirit, charged with patriotism; nation building and fresh processes of union; imperialistic conquests.* It was Dr. Raymond Wheeler's theory that, without exception, governments or states revive from periods of disintegration and civil war on climatic upswings—from the cold-dry to the warm-wet phases of the climatic cycle.

It is during these periods, Dr. Wheeler maintains, that man *rediscovers* the advantages of union, cooperation, and the division of labor. But cooperation is impossible without an increase in governmental control and centralization. Based upon available excess energy—with its resulting optimism, enthusiasm, and aggressiveness—political unities form; imperialism and national feeling rise up; and international wars break out.

During nation-building times, the great international wars are fought with conquest as their aim, and they are conducted with the aid of spontaneous enthusiasm and patriotism of the people and the armies. Among the great nation-building wars of history have been the conquests of Cyrus the Great, in the middle 500s B.C.; the building the great empire in India by Asoka the Great around 280 B.C.; the First, Second, and Third Punic Wars between Rome and Carthage, each of which was fought on three successive peaks between 300 and 150 B.C.; the conquests first of Pompey and then of Julius Caesar; the conquests of Charlemagne, A.D. 780-800; the rule of Alfred the Great in the late 800s; the conquest of England by William the Conqueror, 1066; the conquests of Genghis Khan in the early 1200s; the conquests of Casimir the Great of Poland in the early 1300s, and at the same time the conquests of the Incas in South America; the wars between Louis XIV and Europe in the late 1600s; the Napoleonic Wars and the War of 1812; and, finally, World War I. These illustrate the great outbursts of energy on the main climatic swings from cold to warm, which occur on the average of every 100 years. Climatic, physiological, psychological, economic, and political causes, all integrated in a single pattern, then "conspire" to produce strong and vigorous cultures, states, and kingdoms.

**Climax of the brilliant, Golden Age type of period, and less dramatic and far-reaching cultural revivals that follow the Golden Age patterns; revival of learning and a great outcropping of geniuses in the arts and sciences.* When the upward transition is from a long cold period to a long warm one, a people, nation, or empire, somewhere—frequently several at one time—enters upon what historians have called a Golden Age of cultural enlightenment and achievement, of which strong, but constructive, government is a part. During these brilliant cultural periods arises national spirit in its best spontaneous expression.

Even a cursory examination of the history of science and art will show at once the Golden Ages of scientific, literary, and artistic output.

This chapter can do no more than sketch the outlines of these exciting changes, with brief "pauses for period identification." It will be necessary to show, however, that a Golden Age nearly always arises in each century, just as the world climate swings up from cold to warm. This usually happens about the turn of the century, so that the first quarter is usually what Wheeler calls "the springtime of the century." Also included in Figure 12 are a few of the most famous philosophers, scientists, and literary people of history, such as Plato, Euclid, and Shakespeare. These geniuses prevail during the times of good government. They are symptoms and consequences of the high energy level produced by the favorable climatic conditions which prevail at such times.

Simultaneously, there comes a shift from mechanistic to organismic culture.

Philosophy: rationalism, deduction, idealism; teleology, purposivism, mysticism, "wholism," organismic laws.

Science: organic conceptions of nature, organismic biology, epigenesis; "The plant makes its cells, the cells do not make the plant"; ecology; relativity; search for universal laws; the unity of the entire universe; universal equations; "field theory"; emphasis on pattern and system, coordination, integration. Geometry as opposed to algebra: areas, spaces, masses, as opposed to individual numbers and infinitesimals.

Social science: world views, universal histories, emphasis on cooperation as opposed to competition, emphasis on the state rather than on the individual; socialism, communism.

Psychology: organismic theories, personality as a whole, social adjustment, the subconscious, learning by insight and understanding rather than by repetition; introspection.

The following are some of the more important historical periods that have occurred in this phase of the 100-year cycle:

1. The age of the pyramid builders in Egypt, around 2500 B.C.
2. The age of the temple builders in Egypt, around 1500 B.C.
3. The Golden Age of Greece, between 500 and 450 B.C. (Pericles).

Cycle 2, one of the most famous of all time, extended from around 505 to 420 B.C. and encompassed the centers of two successive cold periods in the 100-year cycle. (In those days the cold periods were short, and the warm periods long. Soon, however, the situation changed, and the warm and cold periods became more nearly the same length.)

On the rise in temperature from cold period to warm period in Cycle 2, there occurred one of the most profound of all the Golden Ages in history.

The Greek city-states blossomed into prosperous communities under the leadership of Athens. One of the basic features of this awakening and upsurge of vitality was a great industrial revolution. With this revolution came a period of unprecedented prosperity. Athens acquired control of the commerce of the Mediterranean and sold her wares, including pottery and products of iron and silver, over a wide area. The Greek colonies became wealthy also, and their size and number were greatly expanded.

Cycle 2 brought with it a worldwide awakening and a worldwide prosperity, comparable on the cycle to the prosperity of the early twentieth century, before the crash of 1929. The worldwide awakening around 500 B.C. included many nations and empires. Persia tried to subdue the Greeks and succeeded in taking some of their colonies at the eastern end of the Mediterranean. China enjoyed a period of awakening, strong government, and prosperity. Empires sprang up in India. Legends point to a period of prosperity in what was then primitive Japan. The Scythians, who were Asiatic peoples, founded an empire. There were strong nations south of the Black Sea, in what is now northern Turkey. There was a so-called Caledonian empire in Scotland.

The Amerindian civilizations, in both North and South America, were prosperous and were enjoying a cultural revival. Egypt staged an important revival. A strong state called Thrace emerged in the area now occupied by Bulgaria and Romania.

The brilliant phase of all this activity was over by 430 B.C. These were "New Deal" times, times of subsidies. It was apparently getting warmer all the time. A Greek architect and town planner, Hippodamus, advocated utopian socialistic schemes. Despotism increased rapidly everywhere, as did atrocities and massacres, just as they increased in Germany and Russia in our own times as it became warmer—and also drier. The birthrate declined as it did in the modern world during the late 1920s and in the 1930s.

4. The Ptolemaic age of Egypt, 200 B.C. Three historically important events must be noted during the boom days of Cycle 4, around 270 B.C. and the years that immediately followed.

One was the emergence of the brilliant Ptolemaic civilization in Egypt, during which Alexandria became a famous center of learning. The second was the rise of the famous Asoka Empire in India, which became a period of unprecedented prosperity for large areas in that country. The third was the rise of Rome to power, for it was during the prosperous phase of this cycle that Rome, with the First Punic War against Carthage, began her foreign conquests.

The records of history point to an era of prosperity in Palestine around 120 to 100 B.C., comparable to the era of prosperity we enjoyed in the late 1940s and early 1950s. This was at the end of Cycle 6. Commerce and industry thrived.

5. Rome at the time of Julius and Augustus Caesar, 60-15 B.C.

A similar era of prosperity was enjoyed during the so-called Augustan age, in Cycle 7.

6. Building of the Eastern Roman Empire (Greek Byzantine) by Justinian around A.D. 527.

Part of the time during the reign of Constantine, at the end of Cycle 11, times were prosperous. Commerce revived over much of the world, but the awakening was most dramatic in the Orient —India, China, Ceylon, and what is now Iraq—and in Africa, especially in Abyssinia and in the Sudan.

7. Founding of Charlemagne's empire, A.D. 771-814.

8. England in the days of Alfred the Great, 871-99.

9. England in the days of William the Conqueror, 1066.

10. The great empire of Genghis Khan, around 1220.

11. The Renaissance and great empire of the Spanish under Charles V, 1500.

12. England in the days of Elizabeth, 1580-1600 (the time of Shakespeare).

13. France under Napoleon and the United States under Washington, Adams, and Jefferson. The Industrial Revolution of 1800 occurred *in exactly the same place* on its own cycle.

Great wealth was acquired in all of the civilized world in the late nineteenth and early twentieth centuries when there was an industrial revolution, based on oil, steel, and the assembly line.

14. The first quarter of the twentieth century: the building of the modern world.

We shall not examine all of the great periods of prosperity of history in this chapter, but they all occurred on the same places in the cycle.

Why all nations or people do not pass equally through a Golden Age on every climatic upswing is an enigma, but Dr. Wheeler felt that his laboriously achieved graphs indicate that *all peoples improve to some extent at this time*, while *showing more aggressiveness* in a political and military fashion. Civilizations too primitive to have a scientific or written literature undoubtedly improve artistically, but it is certain that they improve their economic status—that is, their trade and commerce.

**Growth of industries; shift of population from the country into the cities; rapid growth of cities*. Tariffs are adopted to protect industrial development. There is rapid commercial development; industrial revolutions; a long period of prosperity and rising prices (probably two out of every three years having the properties of a boom); and building of great financial empires.

**Rise of a new aristocracy and a new ruling class; centralization of governments; governments at first democratic and benevolent; economic developments of the private enterprise type promoted.*

Golden Ages—history's best leadership, best governments, and most prosperous times—occur on climatic upswings from the cold to the warm phases of a long climatic cycle, because this type of climatic change furnishes man with a maximum of intellectual and physical vitality. This vitality expresses itself in optimism, enthusiasm, aggressiveness, economic expansion, industrial and social revolution, unification, imperialism, and international wars. The new government, once set up, is at first firm, but constructive, often tolerant. The better judgments and the more difficult achievements are made at this time when man has the energy with which to make them. Unfortunately, of course, when this sort of upswing comes, a tyrannical regime can bring about an age of "iron" rather than one of "gold." Some form of autocracy—absolutism, totalitarianism, or centralization—begins to emerge on the climatic upswing, Wheeler declared, and is in its best phase during the earlier (wet) part of the warm period.

Strong governments continue with a trend toward more and more centralization of power, paving the way for statism; beginning steps are made in the direction of deprivation of the freedoms.

The pattern of warm governments has transposed from one warm period to the next, but the details are different each time. Naturally the character of the political cycle has changed radically in content over the centuries, even though it has kept its form. Gradually the slave classes were reduced, then the serfs, then the proportion of the populations not having the vote; until today the largest proportion in history of the populations of the world shares in some way or another in the change of government. Even in the

1930s, during dictatorship times, large-scale voting was done under constitutions of some kind.

"It is absurd to attempt an interpretation of history on the basis, alone, either of continuity or of discontinuity, for history is both, as in a propagated wave, where a deviation to one side of the axis is still an advance forward, even as the phase is changing," Dr. Wheeler said. "It is the form of the deviation back and forth that has changed but little, if any, . . . but the forward advance of the wave has meant vast changes in the content of the political and military process, as well as in the number of people participating in it directly."

Wheeler did not intend his picture of political cycles to be taken dogmatically, but to represent trends and to point out that the rise and fall, both of centralized and of decentralized governments, is synchronized with the course of the climatic cycle. Even the more primitive warrior nations of the tropics have conformed to the same climatic transitions from the cold to the warm periods. The warm and cold periods of history are worldwide and carry the cycles of rising and falling governments with them. And so it is that during warm times the individual is subordinated to the state, *the part to the whole*; whereas during cold times the state is subordinated to the individual, *the whole to the part*. This situation characterizes the entire culture pattern.

7

The Warm-Dry Phase: From Golden Age to Decay

The trends gradually developing through the preceding warm-wet phase reach a climax of extremes.

A time-honored axiom is that after a good sovereign has ruled—or after a good government has run its course—the succeeding government is not as effective. The trend is always toward decay. Moreover, the trend is from constructive and benevolent rule toward tyranny, cruelty, and decline, toward a less democratic regime.

Culture in general is introverted; there is instability; fifth columnism, fanaticism, surrealism in art, lethargy, lack of individual responsibility.

Severe droughts and excessively high temperatures bring on great economic depressions; failure or collapse of existing economic systems. Also evident is *a period of decline, decadence, and decay in the moral tone of society:* instability of the family pattern; low birthrate; hypersex consciousness; increase of promiscuity and sexual perversions; increase in the divorce rate; growing belief that the state owes the individual a living; shift from interest in freedom to interest in security; less interest in doing a good day's work; loss of interest in perfection of craftsmanship; more pay for less work; communism; socialism; statism; absolutism in government; tyrants, dictatorship; contraction of the freedoms; police systems; spy systems; the individual wants or is forced to live for the state; pogroms; massacres; persecutions; totalitarianism; New Deals; public works; substitution of the state for the church.

And so it turned out that strong governments were typical of warm times. The strongest governments, however, are not always the best. The strongest governments of history are often tyrannical, totalitarian, and either fascistic or communistic. These governments reach their climax during periods of highest temperatures and low rainfall—that is, during the hot-dry periods of history as shown by the shaded parts of the peaks in Figure 12. "Good" governments decline and disappear as it gets warmer and drier. The decline reaches a climax during the hot droughts.

Since Aristotle's time, it has been observed that governments tend to pass through cycles. A democractic government will work well for a time, then it will fall into the hands of a tyrannical bureaucracy from which it declines into a tyrannical plutocracy, or oligarchy, headed by dictatorship. A civil war brings back democracy, and the cycle is repeated.

"It was only on the upswing and during the early part of the warm period that strong governments manifested 'good' qualities such as liberality, constructiveness, benevolence, humanitarianism, foresight, and stability," Wheeler wrote. As the period becomes hotter and drier, this type of government becomes decadent and tyrannical, along with a general cultural breakdown.

Wheeler often said that his purpose was not to evaluate, but to relate the facts as he interpreted them from his studies. He wished "merely to point out the similarity of the pattern throughout history that has prevailed during warm times as it pertains to the relation of the governors to the governed. Since the picture is essentially the same during every warm period, and is always the opposite during very cold periods, climate must have something to do with the condition of the political structure.

"The relationship of government to climatic cycles follows this description very closely, with certain qualifications," Dr. Wheeler wrote in his rough draft for a two-volume study on climate cycles and governments. "The good government that forms on the upswing is at first democratic, because it emerges from a cold period. But at this stage of the cycle democracy does not seem to work for long. The upswing revolutions result in a centralization of the government in the direction of the autocratic type. During the course of the warm period, a new aristocracy arises; wealth increases and becomes concentrated in the hands of a few. This aristocracy collects around the sovereign—or in other ways becomes powerful in the government—even as the sovereign gains in power. In the end, the government is a dictatorship supported by a tyrannical and sympathetic oligarchy.

"All this goes on as the warm period is following its course, until the dictatorship phase coincides with the hot drought, just before the drop in temperature. In the long cycles these developments become extreme; in the short cycles the same trends are still evident."

It is during the warm-dry phase that serious economic breakdowns, epidemics of dictators, and leftist movements come into being. Fascism, totalitarianism, regimentation, and absolutism have reached their climaxes during these periods in history. The government becomes more and more strict, which is an inevitable last phase of the unification process. But this would not happen, or at least would not progress very far, unless it continued warm. As it becomes warmer, the centralization of power accelerates, primarily because of the increased lethargy of the individual under warmer temperatures. Overexpansion and imperialism exhaust the nation just as it becomes warm, dry, and physically debilitating. Economic depression and hot drought, coming consistently together, produce pessimism, lethargy, a willingness—even a demand—for government aid, which means socialism and communism.

The climate now produces social and mental instability, which will lead either to extreme lethargy, complacency, lack of loyalty on the one hand, or to fanaticism and an artificial, explosive patriotism on the other; or it may lead to both, depending upon circumstances. The whole culture pattern becomes sordid and decadent; birthrate collapses; stature declines. The race lacks the energy with which to resist mob emotionality, with which properly to evaluate individual life, even with which to make good judgments. Hence organized intolerance and hatred reach a climax.

Communism, socialism, and despotism, with or without a loyal majority of the populace, are products of high temperatures which reduce the vitality and the aggressiveness of the individual. At these times the individual is willing or able to work only under compulsion. He lacks the spirit and aggressiveness to manage the social order democratically; he must be motivated artificially, as he lacks emotional leadership. The result, of necessity, is either the dole or forced labor under a dictator.

This growth is pushed along under its own gathering momentum, until, as Dr. Wheeler expresses it, the haves separate from the have-nots. Opposed factions or parties emerge—each gathering strength in the general growth process. The haves, the aristocrats and conservatives, become fewer and fewer, but gain more and more in power at the expense of the many. "As the warm period continues, as imperialism increases, and as the state becomes militarized, the reactionary movement becomes absolutistic and totalitarian, whether under a king, a duce, a führer, or a 'dictatorship of the proletariat.' The latter, by the way, is a complete misnomer as far as realities are concerned. There is no such thing as a dictatorship of the proletariat. The only way in which the proletariat has ever 'ruled' at any time in history—and the only way in which it can rule—is through truly democractic instruments."

During such phases, people willingly ask for such political movements. Their vitality level is down; they lack the drive and the will to participate responsibly in a complex society. They blissfully hand more and more responsibility over to their government. Religion becomes corroded in much the same manner. The church goes into decline, and religion becomes little more than a code of social ethics.

Likewise, societies become ill because the individuals within them are not in as good mental or physical condition as they are in other phases of the cycle. Because they lack vitality and stability, men and women do things which, at other times, they would strongly repudiate.

But after a period of lethargy and hardship, it turns cold. A saddle forms. With this change (including the preceding drier years) come hardship and economic recession.

At first there is discontent; but as yet, since it has been too warm, this discontent has been more or less inarticulate and unaggressive. Besides, by now the government has its machinery set up to compel obedience. After it has become cold enough—at or just below the longtime regional mean—man's energy level picks up. The haves and have-nots, constituting a conservative and liberal party from the upper strata of society to the lower, come into conflict. There is civil war. The old order of government is either removed or modified. A new order of government, with reform measures and a new leadership, is in power.

"Thus, when it turns cold, the individual thinks of himself first," Dr. Wheeler observed. "A combination of increased energy, hardship, discontent; an overcentralized and tyrannical government; disgust with growing decadence, spurs him to fight for his individual rights. Here comes the realization that society can be improved only through the work and free expression of the capable individual. . . ."

The "saddle" is not a true cold epoch, Dr. Wheeler was careful to explain. It is only a break in a warm period. Revitalized physically and to some extent freed politically (in the long run the more democratic of the conflicting parties always wins when it is cold, especially on climatic downswings to cold periods), societies take a new lease on life. It turns warm; a reborn national spirit goes on another rampage, and a new outburst of imperialism precipitates more international wars.

Meanwhile, the new party has taken over the machinery of the old government. There has not been time for adequate democratic reforms. The existence of a strong opposing party makes it necessary to maintain a strict government. The growth that began on the first upswing continues. Now, perhaps, the state has a dictator instead of a king. At any rate, it exhausts itself in another series of wars and upsets its internal economy, just before the human energy level is bound to drop. Extreme temperatures and droughts lie just ahead.

"Again a myriad of integrated causes produces a complex result. A letdown in the human energy level comes right at the time when there has been economic overexpansion due, first and foremost, to excess energy; or there has been a devastating war which was due primarily to the same cause, or perhaps both . . .," Dr. Wheeler explained. "Unemployment and droughts require more government action, even to socialistic measures. Economic pessimism, caused both by lassitude from high temperatures and droughts and from the depression, promotes socialism and a willingness, even a demand, that the government take care of the individual. Moral confusion and general decadence add to the necessity of a dictator to keep the state together.

"Meanwhile, any circumstance that occurs to aggravate the discontent of the group as a whole, such as a defeat in a recent war or the presence of a minority group that seems to be getting along better that the group as a whole, becomes a target of the government and the dominant group.

"Unhappiness, together with a general nervous and mental instability caused directly by temperatures that are too high, produces a fanaticism that finds a ready agency for its decadent work in the dictatorship and political machinery that by now have been set up," Dr. Wheeler continued. "Momentum carries the process to an extreme that can be corrected only when the machinery breaks down under the weight of its own corruption and tyranny.

"A practically invariable feature of government in its totalitarian and decadent phase, at the end of a warm period or epoch, was its fanaticism on the one hand and indifference on the other, which applied to the ruler and the ruled alike. Superficially, it would seem that these two qualities are quite contradictory, but on a closer look they turn out to be complementary."

One of Dr. Wheeler's points is that nations, like people, may become physically and mentally ill, grow weak, and break down:

"There are, in general, two categories into which forms of insanity fall. While these are not inclusive, they cover the majority of cases. The one category includes depression, lethargy, seclusion, flight from reality, indifference, lack of emotional tone, schizophrenia, inaction. The other includes elation, overactivity, mania, excessive emotionality, belligerence, and dangerous forms of paranoia. Mental deterioration or decline, then, expresses itself either way, dividing most individuals into these two psychotic groups. On the other hand, the normal individual will fluctuate, under pressure, from one mood—the depressed and indifferent—to the other—the manic and overactive. In an extreme form, either mood is a sign of weakness.

"Societies revealed many of the same characteristics when they become unstable, or went into decline, on the warm side; for it was here that there broke out fanaticism, cruelty, and intolerance as measured by inquisitions, persecutions, pogroms, massacres, and tortures, *all state-promoted*. Either indifference or fanaticism in a people, then, is a sign of weakness.

"It turned out that the more democractic countries or states generally declined through indifference, while the totalitarian and more dynamic states declined through fanaticism. The first political 'psychosis' was more often Western and the latter, Eastern; or, the first characterized the older states that had gone through several cycles: the second, the younger states of more recent unification. In any case, the appearance of these traits was certain indication of an imminent collapse into civil war."

The warm-dry periods in history have included the following:

1. Circa 540 B.C. The Age of Tyrants in the days of ancient Greece (Pisistratus).

2. During the cold period centering on 420 B.C., the friends of democracy attempted to stem the tide of rising despotism; but the cold period did not last long enough. It turned warm again. Warm times breed regimentation and totalitarianism. At 404 B.C., for a year, there occurred the so-called Rule of the Thirty Tyrants.

Along with the growth of absolutism on the warm side of the climatic cycle, various associated phenomena invariably occurred —some form of socialism or communism. State socialism developed hand in hand with Spartan militarism and imperialism, until the family pattern practically disappeared. As the so-called Athenian Empire grew and as the dominance by Athens increased, her institutions became more and more socialistic. State socialism in the ancient world assumed the form of state militarism.

3. 330-320 B.C., a period very similar to the 1930s. Philip of Macedon and his son, Alexander, correspond to Hitler. Fifth columnism was rife in the Greek city-states. The states and empires of the world were all in a condition of decline and were despotic and reactionary. In 323 thousands were banished from Athens by the Macedonians; and in 320 Jerusalem was taken by the Egyptians and thousands were deported.

When the warm phase of this cycle had run its course toward the end of the 300s B.C., a long cold period set in and governments began to collapse. Alexander's empire broke up. Civil wars raged everywhere.

4. 200-190 B.C. In Egypt there was severe oppression. The Macedonians pillaged Greece, burning schools and defacing monuments. The Aetolian League of city-states pillaged Sparta, where democracy was abrogated.

5. A.D. 85-95. Domitian, cruel despot of the Roman Empire; persecutions of the Christians. Hardly had the Roman Republic become a monarchy than the armies determined who should be the emperor, and the man who could control the army controlled the state. Then the state was obliged to care for the discharged legions and their families.

"The growth of socialism and communism as characteristics of collective behavior is inseparably connected with the growth of absolutism and dictatorship," Dr. Wheeler stated. "They grow as the latter grow, become decadent with the latter, and are overthrown with them.

"There is no socialism without authoritarianism, no communism without dictatorship. The one is the necessary complement of the other, for in 'sharing alike,' there is no responsibility of the individual for the individual. When the owner, master, or state provides for all equally, or attempts to do so, then the acceptance of that provision eliminates all individual responsibility for the common good—except on the part of the master. Such has been the character of society, generated by warm periods, whether that society was the Greek state, the Roman Empire, the feudal principality, or the modern communistic state."

6. A.D. 585-600. Mayan civilization in Central America in state of decadence; religion reactionary and stereotyped; priesthood fascistic and "rancid." Fascism in Visigothic Spain. Spain turns Athanasian and burns all literature of the Arian Christians. Justinian's government persecutes all Arian Christians, Jews, and pagans. Cruel and despotic ruler in China. Tyranny and cruelty in the Frankish kingdom.

7. 800-810. Charlemagne's government becomes reactionary. Period of laxity and decline in the Arabian kingdom of Baghdad. China: serious economic depression caused by overissue of bank notes—inflation. Toledo, Spain: chief citizens massacred by despotic ruler.

8. 1000-1030, much of which was warm and dry. Ghazni princes, invading Hindustan, create great carnage; desecration of the temples. Massacre of the Danes in England. Basil, King of the Greek Empire, puts out eyes of fifteen thousand Bulgarians. Jews banished from England. China virtually a socialistic-militaristic state. Land taken over by the state. Hakim of Egypt destroys Christian churches of Syria. France: persecution of the Jews and burning of many so-called heretics at the stake. Poles ravage Germany and carry off ten thousand prisoners.

9. 1255-1270. Kingdom of Castile bankrupt; economic depression; debasing of the coinage; persecution of the Jews. Popular

government abolished in Florence. England: persecution of the Jews. Byzantine Empire: misuse of despotic power. Japan: severe oppression. England: cruel expedition into Wales causes rebellious spirit to last for two hundred years. Italian cities: the doges (city governors) become absolute in power.

There was a bad depression in the 1250s to 1260s during a time when it was mostly warm and dry. This period corresponds to the 1930s. The situation in England at the time was described as "a period of great economic distress." In Castile, one of the Christian provinces of Spain, an economic crisis resulted from a debasement of the coinage. Town finances in France were in a chaotic condition. This was a fascistic and communistic period in history, as in the 1930s. Crops were poor; there were numerous famines; and there were persecutions of the Jews. Governments were despotic, costly, and dissipated.

10. 1365-1375. England: drastic laws under Edward III. Workers could work only at a certain price under penalty of the pillory or the loss of an ear. Noncompliance meant being branded with a "V" for "vagabond." If he attempted to escape, he was branded with an "S" and made a slave for life. If he then protested, he was hanged.

The sequence of events in Europe and other parts of the world in the second half of the 1200s and in the early 1300s is comparable to the sequence of events that the modern world has just passed through and is passing through.

11. 1630s. England: Parliament forbidden to discuss or question any state edict, then dissolved for eleven years. Degenerate phase of the Thirty Years' War: excesses of Wallenstein. Great cruelty exhibited by the Turkish ruler. Poland: persecution of Greek Orthodox Christians. Their priests insulted and churches leased to the Jews. Tyranny in the American colonies. Excesses of Governor John Endecott. Jesuits oppressive in Brazil. Extreme reactionary trend in Japan: anti-European movement; persecution of Christians. Tolerance decreased for all forms of Christianity in England except the official church. Roger Williams goes to Rhode Island to seek freedom.

12. 1670s and 1680s. Fascism and tyranny in France under Louis XIV and in India under Alamgir I. French ravage the Palatinate. In France agents hired to kidnap Protestant children. Protestants could not print books without permission nor sing their own songs on land or water; Protestant churches abolished; many Huguenots flee to England, Holland, and the United States, leaving France in a desperate economic situation. South Carolina adopts Indian slavery. Massachusetts is deprived of its charter. Hungary: Christian priests and teachers seized by the Turks and sold into slavery; thousands flee to Transylvania.

"The absolute monarchy of the seventeenth century was essentially the same in pattern," said Wheeler. "The people were still serfs. The King owned the treasury in the name of the state. The mercantile system was socialistic. State finance from which only the nobility reaped the harvest—when there was any." Ordinarily, a conspicuous warm-dry period develops at the end of the warm phase of the cycle just before the onset of the long cold period. This time the hot drought phase of the cycle was broken up into short stretches, interspersed with revivals of rainfall. As a consequence, it was an unusually active time in history. It was warm and dry in part of the 1680s, and there was a sharp drop at that time in the price of British wheat, indicating the onset of hard times.

An economic recovery accompanied the return of rainfall around 1690. War then broke out between France and England. The American phase of this conflict is known as King William's War.

At the end of this wet period, namely, in the first decade of the 1700s, the price of wheat fell, and again it was warm and dry. It was at this time, in Cycle 25, that a situation developed comparable to World War II. A war broke out between Louis XIV of France and several of the other European countries because Louis wanted to add Spain to his domain.

13. 1930s. Epidemic of dictators worldwide, especially Russia, Germany, and Italy. Russia: communism. Germany: Nazism. Italy: fascism. Many countries changed form of government from democracy to dictatorships. Worldwide devastating economic depression. Leftist movements in old-established democracies like

France, England, and United States. World War II reflecting decadence in society. Cruelties perpetrated in many countries, especially in Germany and Russia. Prisoners of war treated as slaves. Persecution of Jews in Germany. Total war. Willingness to use weapons like the atomic bomb. Seriously lowered moral tone of society. Laxity in sexual behavior. Nudism. Serious rise of the divorce rate. Decadence reflected in art, literature, and music. Extreme temperatures and droughts worldwide. Dust bowls.

Government reached its most deplorable condition during the hot drought period that terminated a warm epoch. The shift toward tyranny was plainly evident even in warm periods of only ten years' duration. Rulers who were benevolent on the upswing always became fascistic if their rule extended into a warm-dry period of any length.

On the other hand, notice that there are names such as Alexander the Great just before 300 B.C.; Diocletian just before A.D. 300; and Stalin, 1930; who belong to the latter parts of the peaks. Still, this is dictator- or nation-falling time. Such times are followed by civil war and the collapse of governments, as temperatures shift to the cold side. We are beginning the twenty-seventh distinct recurrence of this trend in human affairs since the days of ancient Greece.

"Again a combination of causes—economic, political, psychological, biological, and climatic—leads to the next phase in the cycle of events," Wheeler went on. "Before political unity has declined, and while fanatacism is still controlling governmental policy, temperatures start dropping; and the national spirit revives and plays into the hands of a decadent and despotic leadership. Thus imperialism bursts forth once more, and international wars break out on the warm side of the downswing."

According to Wheeler's analysis, wars that come at the end of the drought period have always been the worst wars of history from the standpoint of a disregard for human life. Here is where the majority of the "senseless" wars have been fought; where kings or generals have led their men to the slaughter for a cause that would have been scorned on the upswing; where populations have been disregarded; where noncombatants, women, and children have always been looked upon with the least respect; where the greatest massacres have occurred; where marauding and looting have been at their worst.

More humane attitudes have always been shown on upswing wars. An "international law" of some sort has usually played a part, even though the fighting at that time is usually the hardest.

During downswing wars, there have always been more betrayals, more fifth-column activity, more lethargy on the part of the invaded, more welcoming of the invader, and more dictatorial manipulation of discontented populations.

"All of this results from the fact that, whenever and wherever it is warm for an extended period, the individual becomes less important," Dr. Wheeler commented. "It is then that he is killed with the least compulsion; it is then that fanatic sacrifice for the state reaches its highest pitch under the right combination of circumstances.

"Moreover, the individuals who are now committing crimes against persons in our current crime wave (1950) were born and grew up in the warmest part of the 100-year cycle. At that time, the vitality of the reproductive cells of entire populations had lowered, the birthrate had declined, and it might be expected that a larger percentage of physically and mentally weak or unstable individuals would be produced. It was found that size, in this country, started to go down around 1930, apparently as a consequence of the warm phase—so the experts think, and they are probably right.

"The right kind of properly checked study has not been forthcoming to settle the matter, but during World War II it was reported that the younger generation volunteers and draftees were not as healthy as were men at the outbreak of World War I. This statement has been denied, but it would not be at all surprising if . . . true. We do know that during the warm late 1920s and the 1930s, general vitality declined both in human beings and in animals.

"The climatic factor, then, has undoubtedly intensified the moral breakdown (taking the population as a whole) that followed the war by producing a crop of individuals of less than normal stability. This would increase the number of criminals in the population.

"The vandalism of early teen-agers is correlated, we believe, with the fact that they were born during the climax of the heat wave in the middle 1930s, when the vitality of the population was at its lowest and the birthrate was at the bottom of its decline. It will be interesting to watch what happens in this age group five years from now when they will be in the neighborhood of twenty. Shall we witness an increase in teen-age crime on their account? Will they be more unstable, on the average, than others were at the same age?

"Given a background of instability as a result of the weather, the younger generation is now living during a run of years that is excessively stimulating and that generates unusual amounts of human aggressiveness and desire for action. World War II, the great boom, and now the cold war and the numerous revolutions and civil wars over the world are symptoms of this stimulating drop in temperature.

"The change in the weather trend, however, stimulated the stable and the unstable alike. The unstable have been unable to harness their greater aggressivenss into constructive channels; hence the timing of the crime wave. Thus, the predisposition to crime in the first place, and the incentive to commit it in the second place, were both conditioned . . . by the longtime weather trends.

"Now, the same weather trend that has presumably predisposed these people toward crime helped produce and intensify the great depression of the 1930s. Most important of all, it timed its occurrence. Then the opposite weather trend—cold-wet instead of warm-dry—that stimulated the crime wave also stimulated World War II and the accompanying and subsequent boom. Criminals, then, are more aggressive at the same time that businessmen are, because of the prevailing weather trends."

8

The Cold-Wet Phase: Rising Tides of Civil Strife

Thus it would appear in the vast research materials assembled by Dr. Wheeler that absolutism, socialism, communism, and totalitarianism, on the one hand, and democracy, individualism, and laissez-faire, on the other—the two opposite types of social organization—are definitely conditioned by opposite types of climate. A democracy will not last through or develop during a warm period. Absolutism, tyranny, socialism, and communism will not last or develop through a cold period.

But in keeping with their cyclical nature, temperatures continue to drop. It becomes more stormy and more envigorating. The mounting discontent with the government, plus a possible economic exhaustion or defeat in an international war, or both, plus the continued drop in temperature, lead to the next phase of events. But since this phase occurs at no other time than with a drop in temperature, the stimulating, but also irritating, influence of a changing climate is an essential factor in the picture. A temporarily increasing rainfall, and decreasing temperature, bring personal discomfort, because the winters are long and cold, with many storms, or perhaps the droughts are continuing.

While still warm, but under lowered temperatures and increased rainfall, we find an outbreak of international wars, usually promoted by the dictator nations (less democratic and more fanatic), and reflecting the decadent condition of society in general; called nation-falling wars because governments are in the course of collapsing; more cruel than the nation-building wars; "total war"; prisoners and populations either slaughtered or enslaved; climax of lying and trickery; climax of graft; cultural thefts; climax of pogroms and persecutions of dissenters; race antagonism; ruthless methods of fighting; disregard for efforts to control war; worthlessness of promises and treaties. It became evident to Dr. Wheeler that governments gradually decayed, in one manner or another, as the warm period continued; but that little or nothing was ever done to prevent this development *until it turned cold.*

Also characteristic of cold-wet times are:

- Appearance of large sunspots and a temporary shortening of the sunspot cycle, presaging the shift from the warm phase to the cold. Outbreak of aurora, of which there have been only a few visible at temperate latitudes during the previous warm phase; lowering temperatures and increasing rainfall; beginning of a second period of storms and floods.
- Sudden and marked increase in the birthrate; period of high energy level; great improvement in health in population at large; general increase in alertness, initiative, and aggressiveness of the individual in all areas of human action and achievement; betterment in the moral tone of society.
- Migrations from the cities to the country; a return to the land; shifting of homes and industry to the country and to undeveloped areas. Beginning of migration from one country to another; cultural intermingling and borrowing.
- Revival of foreign trade and investments; increase in travel for both

business and pleasure; period of inflation; high prices; prosperity; maximum in business volume; booms. Business boom periods also come at the end of a warm cycle when temperatures are falling and the cold-wet cycle is beginning. Another extended period of prosperity and high prices begins, but in general the price trend is downward. As the cold period advances, human energy again decreases, and another depression occurs before it starts turning warm. (We are in such a period in 1975.)

- Decline and decadence of the old aristocracy and ruling class; reactionism whether the government is an absolute monarchy, a republic, or a socialistic dictatorship; decentralizing trends in government, business, the church, and in other basic human institutions; antitrust movements. Great religious leaders who advocated the dignity and importance of the individual are typical of cold times. Jesus lived during cold times, as did Confucius and Mohammed. A period of emancipation, democratic reforms, and revolutions; new and more liberal constitutions; inauguration of constitutional government; broadening of the franchise; civil rights legislation, freeing of serfs, slaves, and underprivileged classes; removal of restrictions; overthrow of despotic governments and dictators; climax of management-labor problems; strikes and other labor difficulties; liberalization of laws and religious creeds; reform movements in religion; return of laissez-faire thinking and policies; revival of private enterprise; reaction against socialistic and fascistic trends; antistatism.
- Termination of international wars and beginning of a long period of rebellions, revolts, insurrections, *coups d'état,* palace revolutions and murders, party struggles, factionalism, sectionalism, religious controversies and wars, race riots; vigilantes, assassins, Ku Klux Klans, and secret societies. There are sometimes great wars fought on the falling side of the curve, at nation-crumbling time, but these lack the enthusiasm of nation-building wars and are often successful only with the aid of fifth columns. The stage is set for rebellions and civil wars. The old national spirit, the old loyalties, are rapidly disappearing. There is disillusionment. It is decided that the hopes entertained in the previous outburst of nationalism were false; the masses are bitter. Anarchy increases.

"Consistently," Wheeler wrote, "during the better phase of democratic movements, tolerance, Christianity, interest in education, cooperativeness, enthusiasm for the individual, and orderliness prevailed. There was much travel, cultural borrowing, and an extension of trade. The breakdown took the form of anarchy, intolerance both of outsiders and of one another, race and class riots, assassinations, and sabotage. During cold times the government usually attempted to control the persecutions of racial or other minorities; but during the warm droughts, these have practically always been government-sanctioned or government-promoted events.

"Just as the antecedents of political communism were serfdom and slavery, so the antecedents of democracy were piracy and anarchy. Societies during the cold periods of ancient times were in a state of complete disruption. There was little or no central government. Populations were at a minimum. Many cities were empty, deserted, or in ruins. Marauding bands and migrating hordes ravaged the countryside and pillaged the towns. The seas were infested with pirates.

"The next step was taken when depopulated civilizations were able to survive under some degree of order during cold times by maintaining trade routes and colonies and mercenary soldiers as protectors of commerce. Towns were protected by garrisons to ward off robber barons and guerrilla bands. Knights in castles, surrounded by their serfs, protected themselves and their vassals as best they could against one another and wandering tribes. Commerce was hindered by thousands of local tariff barriers and tolls; profits were reduced by the necessity of paying tributes for being left unmolested. Each community had its own laws, coins, weights, and measures. Life was insecure at best. Rival cities or small principalities fought one another over their commerce until it became apparent to the wiser among them that prosperity could not exist without order or justice.

"When state governments were in the early stages of their development and standing armies had been organized, these armies frequently broke loose during cold times and became pillagers and marauders, headed even by members of the nobility. Secret societies of assassins, ex-soldiers, and adventurers carried on an unhindered piracy on land, while sea pirates patrolled the seas. Small gangs formed that were led by a 'Jesse James' who preyed upon commerce and farms. Political parties fought wars of annihilation; among the nobility, feuds were common that hurled loyal followers into senseless bloodshed.

"Older states split into warring political factions during cold times, frequently with opposing armies. Subordinate political units seceded or fought wars of independence, while political parties wrangled and hired gangsters to kill off each other's leaders.

"When the state became firmly established and its economy fixed, then, during cold times, business insisted on free competition. Ruthless, unfair practices were countenanced, and economic wars that annihilated the 'little fellow' invariably began. The attitude was 'everybody for himself ' and a minimum of regulation by the state. Spoils systems and graft dominated the relationship between business and government.

"In short, there has been a pattern on the cold side that has transposed from one cold period to another down through history, a pattern whose extreme form has been anarchy pure and simple, ranging from wars, intrigue, and treachery among the governors and their loyal followers to commercial war, race and religious riots, and armed civil war among the governed. All this is the *fanatic* aspect of cold times. The 'lethargic' aspect has always assumed the form of neglect, debauchery, and extravagance on the part of rulers and the upper classes, and listlessness, pauperism, begging, itineracy, license, rapine, and vagabondage among the lower classes."

After a succession of serious droughts in the center of the cold period, the rains begin. The rains, together with cool temperatures (the midline represents temperatures in the temperate zones of around 50° F.), raise the vitality of the people of all classes to its highest level, just as the curve, as drawn, reaches the midline on its upward swing. It is raining most at the midline.

"Civil wars, fought after the tyranny has become intolerable, and the state is disorganized, coincide with the drop of temperature into a cold period. These civil wars result in democracy—or at least in democratic reforms. In the extreme case the condition in some instances is anarchy, out of which democracy in its best state emerges. This may come during the earlier part of a cold period, but generally toward the end."

Sometimes the governments are able to sustain themselves for a while, but eventually they will fall—if it stays cold. In spite of democratic reforms, leadership is generally lacking and frequently corrupt. Meanwhile, as one might expect during a period in which the individual life is rediscovered, society becomes flippant and none too orderly. This is the true Romantic phase of history, with its appropriate literature, art, and costume emphasizing the individual's rights, feelings, and self-expression. Thus, there is a reaction against the old order, against the old aristocracy and everything it produced. Culture becomes superficial, but out of it a new order emerges, a new commerce, a new leadership of better quality emerging from the rebellious and freed elements of society.

Sovereigns during cold civil war times are not likely to rule very long. They are often assassinated or run out of office after being in power a short time. Governments change hands rapidly during cold times. Even during the long reign of Queen Victoria during the cold mid-nineteenth century, party government changed hands numerous times.

There are, of course, leaders noted for their constructive work during civil war times. These leaders promote democratic forms of government, or they are patriots who lead the struggle for civil rights. Such were the Gracchi in Rome, shortly after 200 B.C.; Marius, shortly after 100 B.C.; Hus, the fighter for religious freedom in Bohemia, right after A.D. 1400; Cromwell in England in the middle 1600s; and Abraham Lincoln in the United States in the middle 1800s. But there are not nearly as many of these earnest reformers as there are good leaders of nation-building time on the upward slant of the curve.

A strong, far-seeing, and democratically minded ruler has at times been able to pilot the state through the downswing, through the civil war epoch, and been able to give his people the needed democratic reforms. Certain Norwegian and Danish rulers seemed to have been able to do this. Peter the Great combined the qualities of a tyrant and reformer, on the downswing, and probably kept Russia free from anarchy. Lincoln gave promise of being a good downswing leader; Jackson was not.

The following are famous cold-wet periods in history:

1. 320-310 B.C. Alexander the Great, at nation-falling time in the late 300s B.C., carved out a shaky empire which, with his death, collapsed into civil war. The famous Chandragupta expels the Greeks from India and paves the way for a recovery of India.

2. 140-130 B.C. Jerusalem under Simon the High Priest; peace and economic prosperity. Rome: revolutions of the Gracchi in interest of democratic reform.

3. A.D. 0-30, which contained many cold-wet years. Revolts worldwide. Teachings of Christ with emphasis on worthiness of the individual life and in democracy.

4. Conquests by Attila the Hun in the first half of the fifth century A.D. contributed to the fall of Rome.

As Raymond Wheeler has told us, there are regularly two main periods of prosperity in each of the 100-year cycles. One comes on the upswing of temperature from the cold to the warm phase, the other occurs on the downswing of temperature from warm to cold, where we were in the late 1940s. But the documents of the past covering the second part of the cycle are not as complete as those covering the first part, because governments were crumbling and about to collapse. Revolutions, insurrections, and civil wars were brewing. These are not times favorable to the making of permanent records. On the other hand, during the swing from cold to warm, governments are strong and better records are kept. By the Middle Ages, however, there exists more detailed information and indisputable evidence that periods of prosperity, such as the one we have just left, have been regular occurrences down through history in their places on the cycle.

On the cycle a thousand years ago we find a time sequence comparable to where we are now. It was at the beginning of a long cold phase in Cycle 17, and the first weavers and fullers (cloth thickeners) were established in the Belgian city of Ghent. The textile industry expanded rapidly in western Europe. Private enterprise was encouraged by the King of Wessex, in England, and by rulers elsewhere. Merchants began a long struggle for freedom and power against the nobility, which they eventually won. Feudalism was becoming evident as the coming economic and political pattern of the later Middle Ages.

These were extremely dynamic times, and the civilized world was seething with change. The Norsemen of Scandinavia and the Slavs of Bohemia were beginning to embrace Christianity. The Norse and Slavs were fusing in Russia. The Saxons and Danes were blending in England. A new world was being born and an old world was dying.

The new system, feudalism, was to last five hundred years. In Japan the provincial military class of society sprang up. Chinese classics were printed for the first time by means of wooden blocks as a cheap substitute for stone engraving. The oldest continuous parliamentary body in existence was established in Iceland. There was a great awakening in Germany under Otto the Great, which marked the beginning of the Holy Roman Empire. All these dramatic rebirths and beginnings are comparable to the awakening that is going on now.

Exploration and migration dominated the scene. Civil wars were fought everywhere. There were violent disputes between the monks and the clergy, between emperors and popes, between the new and old classes of the nobility, between invading hordes and the native populations. China hurled back hordes of Tartars bearing down from the northwest. Norse and Russians sailed down the Dnieper River to the Black Sea. The Arabs invaded India.

5. 1035-1045. Ferdinand the Great unites León and Castile after subduing the rebellious barons. Henry III: Holy Roman Empire at height of its power; era of great prosperity in the towns. Truce of God declared by the Pope to keep down private wars. Norway: Magnus the Good, an outstanding ruler.

6. 1100-1105. England: Henry I grants Charter of Liberties, removing abuses of previous tyrannical government. Milan declares herself a republic.

7. 1134-1140. Aragon: representatives of the cities sum-

moned to the assemblies (Cortes), an important step in the evolution of democratic government. Holy Roman Empire: power transferred from the Emperor to the Diet. France: Louis VII supports the communal movement, a movement in local self-government.

8. 1130-1175, containing many cold-wet years. Denmark: Waldemar the Great, a prosperous reign; one of the outstanding rulers of history; democratic. Beginning of the Bank of Venice. England: the great Council and the Constitution of Clarendon, secular control over clerical courts. The Assize of Clarendon: reform of the judicial system; trial by jury; beginning of the grand- and petit-jury systems; recognition of the common law; circuit judges. Egypt: Saladin, extraordinarily good ruler; treated Christians magnanimously; symbol of Saracen chivalry; sympathetic, gentle, and devout. Venice becomes a great maritime power.

9. 1215. Strife between the barons and King John of England. The Magna Charta.

10. 1280-1290. As a result of the depressions, tyranny, and finally, the falling temperatures, a great period of democratic reform took place. In the 1280s and 1290s, *approximately where we are now on the cycle,* the Magna Charta was reaffirmed after having been repudiated. Model parliaments were instituted in England and France. Numerous legal reforms were achieved. In 1283 a Magna Charta was exacted from the King of Aragon in Spain. The famous Italian poet Dante was instrumental in effecting democratic reforms in Florence.

Sweden: Magnus the Good. Portugal: progress and economic prosperity. England: the famous model Parliament. Economic troubles began in the middle and late 1280s and lasted until after 1300. These particular hard times came during a period, generally, of great democratic reforms. They are comparable to the depressions in the modern world during the mid-nineteenth century, which was also a time of great democratic reform. They are comparable to the depressions that occurred in the late 1950s, the late 1960s, and which lie ahead in the 1970s.

11. Around 1420-1430. The Hussite Wars fought for religious freedom, anticipating the Reformation. Poland: the Polish habeas corpus—citizens cannot be imprisoned without trial.

12. 1640s. England: the famous war between Parliament and the Crown (Cromwell).

13. 1680s. The famous British and other revolutions. Centering on 1710, there was a strong boom in the same place on the cycle where we were in 1952. Temperatures had fallen, shifting back and forth across the midline between the warm and the cold side as temperatures did in the late 1940s and early 1950s. A brief cold-wet phase developed. Speculation went wild. There was a strong revival of private enterprise. The "socialistic" state system of economy and mercantilism had run its course under the despotic rule of Louis XIV and other totalitarians. The French state was becoming weaker and weaker. In Turkey the Shah handed over the government to his wife, who did as she pleased while he lived a life of voluptuous pleasure. Louis XV came to the throne of France, but he was indifferent to the welfare of the country. In Poland the government reached such a low ebb that no work was done and no laws passed in two generations because of the unscrupulous youth of the country who predominated in the Diet and ruled with absolute power. So it went around the world; but nevertheless, it was a time of temporary prosperity, while governments changed hands. The price of British wheat jumped from an index of around 70 to almost 200 during this boom and wet period. Remember, this is comparable to where we were in the 1950s. Numerous new business adventures and speculations were undertaken; and, for a while, they prospered.

14. 1810s. Revolutions in Latin America leading to independence.

15. 1830s. Revolts and rebellions universal. France: revolution in Paris. United States: Jacksonian movement. California and Texas revolt against Mexico. Greece and Holland independent. Slavery abolished in British colonies. Reform bills in England.

16. The 1860s. Great civil war in the United States. Other civil wars in Poland, Italy, Japan, and elsewhere.

17. The 1880s. Long list of constitutional reforms and liberalized governments.

18. The late 1940s. Economic prosperity in many countries of the world; rapid economic recovery in war-devastated countries;

inflation. Widespread labor strikes. Civil strife in Palestine, India, Greece, Burma, Java, and China, and troubles in Italy, France, and many Latin American countries. Increasing resistance to Russian tyranny in the Balkans. Signs of increasing internal dissatisfaction in Russia. Dissatisfaction growing in England toward leftist regime. Resistance growing in the United States toward leftist trends.

All of these events and occurrences reached a speed and intensity which happens only once in five hundred years—in the 400s, the 900s, the 1400s, and now, the 1900s. History is repeating itself at a higher level of technological, economic, and political progress than ever before; but, nevertheless, it is repeating itself.

The conquests of Tamerlane in the late 1300s; the last wars of Louis XIV in the early 1700s; and World War II in which Hitler, like Alexander the Great, attempted to carve out an empire, come at nation-crumbling time. We are still in such a period. All governments that did not change after World War II will do so not far in the future.

The three major events of the political cycle—1) the strong, constructive government of the upswing; 2) the despotic and decadent government of warm-dry times; 3) the decentralized, often "anarchistic," but democratic government of cold times—are products of the relative amounts of energy which man prossesses for achievement, and of the sequence of high and low energy levels. These levels are determined in the main by climatic conditions.

According to Dr. Wheeler, "Three major facts pertain to the rise and fall of governments, that, all through the investigation, were so invariable and their relationship to climate so precise as to challenge an attempt at an explanation in general terms.

"*First*, there were the occurrence of Golden Ages, the rise of strong governments under superior leaders, the outburst of imperialism and international wars on climatic upswings from long cold periods into the warm-wet phase of the climatic cycle. *Second,* the decline, onset of decadence, the growing excesses of centralized government, the emergence of dictators, tyranny, fanaticism, communism, and socialism, as the warm epoch continued, and as temperature and dryness increased. *Third,* the occurrence of civil wars, rebellions, and revolutions, the origin and growth of democratic institutions and individualism, during cold periods. . . . No law of chance can explain the fact that undemocratic trends are invariably associated with the warmer climatic phases, and democratic trends with the colder phases. No law of chance can explain why international wars so consistently predominate on the warm side and civil wars on the cold. Relationships so consistent, universal, and precise point directly to a causal factor or set of causal factors."

9

The Cold-Dry Phase: From Anarchy to Democracy

Some form of increased democracy–individualism, liberty, or laissez-faire —emerges on the climatic downswing, and is at first anarchistic in the sense of civil war and its aftermath. Then for a time democracy is at its best, especially if good leadership is available. This is generally during the later part of the cold period near the upswing, provided the upswing is not too dry and warm. But the more democratic type of government is inclined to decay during the cold-dry period.

"Democracy declines when it is cold and dry, but as in the case of the decadence on the warm side, nothing is done about it until temperatures begin to . . . approach the longtime mean when it becomes wetter and more stormy. In this new outburst of rebellions, the people clamor for better government. The chaos of revolution

Figure 13 THE 100-YEAR CYCLE

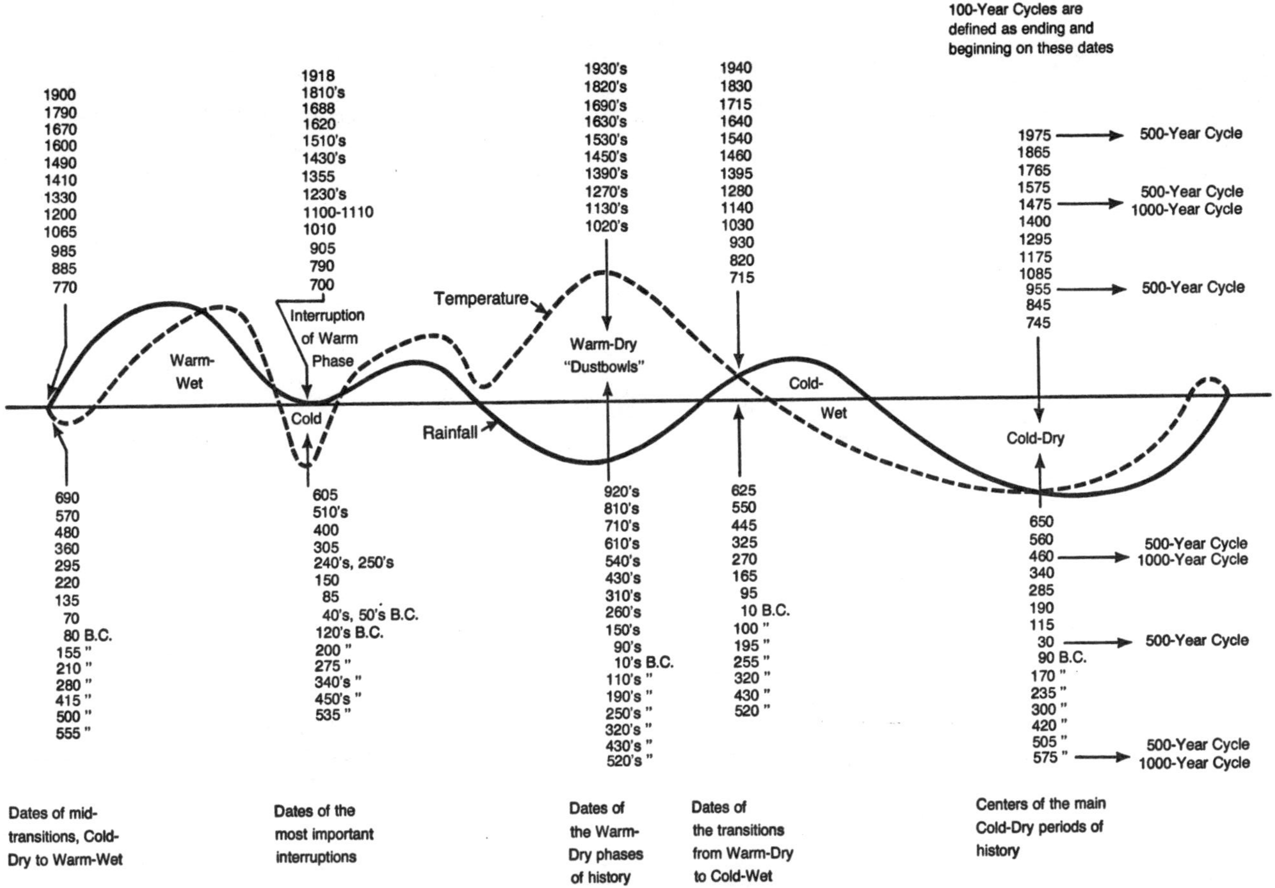

ends in autocracy, while the chaos resulting from the downswing civil wars ends in anarchy."

The data on weather-energy cycles continually indicated to Dr. Wheeler that after a dull period during the cold droughts, when democracy becomes decadent and the spirit of everyone for himself has led to an intolerable state of corrupt and wasteful affairs, climate begins its climb to the next warm epoch. Democratic institutions emerge from a cold period because then the individual's discontent is backed by a sufficient amount of aggressiveness, individual spontaneity, interest, and optimism to promote the fighting spirit necessary for the achievement and maintenance of individual freedom in any organized society. Despots are no longer tolerated, because the invigorated individual thinks of himself and his rights first, and of social welfare second. When the individual is lazy and incompetent, he thinks of the state first, because he is then willing to be fed or then wants the state to take care of him.

If, at such a time, conditions promote a secondary revival of national spirit, as they will do on a secondary upswing following a saddle period, the willing subordination of the individual to the state plays directly into the hands of fanaticism; and the political structure goes to absolutistic or communistic extremes. From the standpoint of individual freedom, the one type of government is the same as the other. Both are totalitarian in political control and authority. Both may be totalitarian in the temporary loyalty of the subordinated subjects.

Of course, as Dr. Wheeler observed, Dark Ages and periods of superficial culture have always occurred during cold periods. But there was a gain as well that transcended the loss; for after each cold period civilization blossomed forth at a higher level than during the previous warm period, just as in the case of animal and plant life through geologic time.

"The Golden Ages of history, the best in human health and leadership, cultural output, the great periods of economic and political growth and expansion, have occurred after a toughening process has been going on that has revitalized the race at the biological level," Dr. Wheeler writes. "Moreover, during cold times cultures came in contact with one another during migrations, travel, exploration, and colonization—all of which extended to some extent into the earlier part of the warm period."

"In the hands of a new generation, a fresh national spirit wells up, and revolts occur against frustration," Dr. Wheeler asserted. "Enthusiasm, optimism, and aggressiveness, organized through a social revolution, result in a new state . . . . As democratic government continues, it tends to become bureaucratic—either in the hands of leftovers from the previously dominant aristocracy . . . or in the hands of a new generation of rulers who have come into power through intrigue, wealth, or some other form of leverage. A new set of rebellions break out following the dry years on the cold side; and during these rebellions, effort is made to overcome the evils of decadence in the democratic pattern, or the tyranny left over from previous warm times. . . .

"A strong leader comes to the front. A new Golden Age is on, and a new cycle of imperialism begins. The revolutions result at first in democratic reforms, because they begin on the cold side. Were it to remain cold, these reforms would remain; but as it becomes warmer, the more power the 'radical' party assumes.

"After a reign of terror, the new spirit coalesces into a strong, centralized government that, from the standpoint of individual rights, is reactionary."

Cold-dry conditions include:

- Long droughts during years of low average temperatures (conditions that did much to stir up the great migrations of the past).
- Continuation of the cold-wet trends.
- Climax of migrations and immigration. Climax of cultural intermingling and borrowing. Democratic trends go to an extreme; a climax of anarchy, piracy, vigilantes, Ku Klux Klans; democracy out of control; general weakness in government; climax of class struggles and racial antagonisms; climax of religious controversies.
- Increase in foreign trade continues; emphasis on free trade. Long period of economic depressions and low prices; business good in about one out of every three years.
- Era of profligates, spendthrifts, and immoral rulers and sovereigns.

Tendency for another period of decadence to develop, this time within the mechanistic, democratic pattern as opposed to the organismic, autocratic pattern.

- Shift from the stadium principle in recreation and in athletics to the Olympic principle, emphasizing more participation of the individual. Revival of religion; evangelism as opposed to social ethics and abstract religious philosophy; revival of emphasis on the individual conscience and the relationship of the individual to his God; Sunday school movements; era of hymn writing and religious music to be sung by the people, as opposed to anthems and masses to be listened to.
- Feminism; labor movements; international workers' associations and like organizations. All through the cold period, a general awakening of the "proletariat," or masses of the people; rapid improvement in intelligence and knowledge; spread of educational institutions; increase in free public education; gradual reemergence of national feeling; revival of folklore; wider use of local languages as opposed to the language of the aristocratic class or of the country to which the people have hitherto been subject; conquered countries and races acquiring freedom.
- At the end of the cold period after the droughts are over and rainfall has picked up, and with temperatures rising, a fresh outbreak of revolutions set up new and strong governments which in one way or another are always more democratic than any previous governments; final overthrow of the old tyrannies; the new governments, benevolent and democratic as most of them are, initiate the process of restoring regimentation and tyranny as the cycle in government repeats itself; the period of nation building again, the period of Golden Ages and the Renaissance phase of the culture cycle; soon an outbreak of international wars as temperatures approach their longtime averages and cross to the warm side.

The picture, as Wheeler described it, holds best for times when events occur more or less by themselves, without the appearance of unusually good rulers or statesmen to guide the ship of state. Modifications of this pattern to occur when the state is guided by strong hands or by a ruler with unusual character and insight. Such a ruler, on the upswing, may have come into power several years before it turned warm and given the people a democratic government which he was able to preserve through the cultural changes of the upswing, indeed, until his death. However, Wheeler found, such a ruler invariably experienced difficulties at the hands of his subordinates, who became more autocratic when it turned warm.

Some of the famous cold-dry periods of history are as follows:

1. 310-290 B.C. Furious civil war and anarchy in the empire of Alexander the Great. Migrations and piracy.

2. 185-155 B.C. Sparta corrupt, self-indulgent, without loyalty or religion even down to the swineherd; class wars; finally looked to Roman yoke with relief. Syrian bands overrun Palestine; try to make the Jews worship heathen gods; religious wars among the Jews; Maccabeus finally expels the Syrians.

3. A.D. 50-70. Rome: Nero; palace crimes; short-lived emperors; Jewish revolts.

4. 170-200. Anarchy the world over. Rome: state of disintegration; political anarchy; short reigns and palace murders. Numerous migrations over Europe and Asia. Piracy.

5. 445-480. Downfall of the ancient world everywhere, including fall of Rome. Attila's great empire falls apart. The Vandals sack Rome. Anarchy and religious wars in China. Strife for supremacy among the bishops of the church. Romans completely helpless to control their affairs. Invasions of England by the Jutes. Huns, Gepidae, Goths, Langobards, and others roam central Europe at will. Scythian Tartars invade northern India.

6. 630-690, most of which was cold and dry, but with a considerable number of cold-wet years. Religious wars among the Arabs and victory of Mohammed. Spreading of the Arabs while engaged in bloody civil wars among themselves. Period of "do nothing" kings in the Frankish kingdom. Anarchy worldwide.

7. 840-880. Violent civil wars everywhere. Charlemagne's empire crumbles and is divided. Germination, however, of many new kingdoms later to become European states—England, Norway, Holland, Germany, Scotland, Bohemia. Beginning of the feudal system. Piracy. Numerous migrations. China: violent religious wars. Norway: many nobles take to piracy. Norsemen overrun Europe. Beginning of Russia.

8. 930-985. Great social and economic changes as feudalism is

established. End of first half of the Middle Ages, terminating a 500-year cycle. Anarchy, piracy and civil war universal. Spread of Christianity and new forms of private enterprise, as for example, first weavers and fullers established at Ghent in Belgium. Ethelstan, King of Wessex, encourages private enterprise. Corruption in the Church.

9. 1460-1490. Fall of the Middle Ages. Great climax of civil wars. England: Wars of the Roses; end of feudalism in Britain and beginning of the party system. Life in Florence described a time of people growing rich while being solaced by shows; dilettantism and empty learning; paganistic and atheistic boastings; a proletarian era. Bakers in Swabia defied several imperial cities in a sort of labor strike. People everywhere cruel and arrogant. Japan: the peasant class almost disappeared as a result of the long civil wars. Civilizations in the Western Hemisphere in state of disruption. Status of Holy Roman Emperor reduced to nothing. England: Edward V, thirteen years old, and his younger brother, the Duke of York, confined to the Tower of London and cruelly murdered.

10. 1715-1745. Jacobite revolt. In the late 1710s it turned cold and dry, just as it began to do in the first half of the 1960s. A French investment and real estate promotion known as the Mississippi Scheme—an early example of modern speculative finance—suddenly collapsed, bringing widespread panic and disaster. At the same time a British speculation scheme known as the South Sea Bubble burst, causing a panic in England. It was not accidental that these two booms occurred at the same time and collapsed at the same time. The period of prosperity and speculation was as wide as civilization—as was the resultant economic decline. Of course in some places it was worse than in others. Trade declined between Spain and Latin America. Centered on 1720 the British wheat price hit another low. Civil wars in Persia, Turkey, Hungary, China, Russia, Ceylon, Mongolia, Morocco, Bulgaria, Tibet, North Africa, and the Philippines. Tensions beginning to develop that later were to crop out in the American and French revolutions. Wave of materialism and skepticism.

11. 1780s. Difficulty in American colonies following the rev-

olutions; indifference toward union; isolationist attitude among the colonies. Tension rapidly mounting in France; the French Revolution. Civil strife worldwide.

12. Late 1880s and 1890s. Anarchist movements in Europe and the United States. Revolts universal, but especially in Cuba, Philippines, Russia, Chile, Argentina, Egypt, South Africa, Brazil, and Panama. Reform measures. Civil troubles in France, Russia, Romania, Belgium, Germany, Ireland, Armenia, and Turkey. Labor movements universal. The "Gay Nineties" illustrate the loutishness, dilettantism, ostentation, glitter, pomposity, garishness, theatricalness, and jingoism of cold-dry periods.

Raymond Wheeler was to note that in locating these different cultural events in relation to the climate curve, "exceptions" will be found. That is, an event that is typically warm may appear a little ahead of time or may last over into the beginning of the cold period.

Where exceptions occur, the problem is to understand why they occur. Since the climate curve is a generalized one representing the world as a whole, and since the different areas of the earth do not shift from the warm to the cold or cold to the warm side at *exactly* the same time, not every item will fit unerringly into the phase of the climatic cycle in which it might be expected.

A discrepancy of this sort might mean that the climatic phase was a few years ahead or behind time in the section of the world where the event took place. Originally, Wheeler plotted the data separately by countries; but there was so close an agreement on the timing of the events—and the problem was so obviously a world problem as opposed to a local one—that the results were charted together.

A discrepancy might also mean that local conditions, over and above the weather trends, had sensitized scientists, artists, political leaders—or even the population of the area as a whole—in such a manner that they responded more quickly, even to slight changes in the weather trends, than did the populations in other countries. At almost any time there are individuals, especially in science or in art, who anticipate a universal trend soon to follow.

Or a discrepancy might mean that local and man-made conditions are causing a lag in the timing of events by climatic factors. Human forces are operating against the climatic. Social, economic, and political forces in their own right, or the personality of a particular leader, may cut across the influence of the longtime weather trends and alter the usual timing of the event by the weather factors. Social and political institutions tend to perpetuate themselves by means of their own inertia. A given event or the work of a given individual is often "transitional" in character: it is partly "warm" and partly "cold." In many instances, a given person or group outlives a particular warm or cold period and, although generally under difficulties, continues on within the pattern of the previous phase.

"The surprising fact is that the vast majority of historical and cultural events, regardless of location, are timed, all down through history, with the prevailing weather trends," Wheeler stated, assured by the demonstrable evidence of his compiled data.

Notice on the chart how regularly all this has happened, how evenly the black dots are spaced. The 100-year cycle may contract and expand at times (90 to 110 years), but it averages close to 100 years and has maintained that central figure down through history. There is every reason to suspect that the cycle will continue to behave in this fashion. There has not been an extended cold-dry period since the nineteenth century, but we have just now entered such a time, which shall continue to the year 2000.

As Wheeler put it: "Judging from the behavior of plant and animal life during geologic time, the de-differentiation, or general simplification, of plant and animal forms during colder periods was in itself a factor connected with the luxuriant outbursts on the next long upswing into a warm-wet epoch. Plainly, life was at its maximum vigor just as it became warm, or shortly thereafter. This was because life had been invigorated and strengthened during cooler temperatures; for then, of necessity, life must have been hardier—so long as it did not become too cold. This hardiness blossomed into maximum fruition on the next climatic upswing." A kind of spring for life in general.

10

The Warm-Wet Arts

Dr. Wheeler's contention that even man's spiritual-artistic expression is subject to the weather cycles may provoke hostile mutterings among those artists who stoutly maintain that they are fiercely independent creators who remain aloof from the ebb and flow of external forces.

"A concept of weather as the major influence upon our lives . . . might at first seem to resemble the astrology of a more mystic age, with the control of man's destinies determined by the stars," Dr. Cleta Margaret Olmstead, Associate Professor of Art, Florida State University, wrote in an article for Wheeler's *Journal of Human Ecology*. "Yet anyone who has changed his place of residence, even in these United States, has discovered both in the society and in his surroundings, as well as in himself, a radical difference in appearance, attitude, aptitude, and achievement. . . . That we are affected by seasonal changes, the spring of growth, the summer of fruition, the autumn of decay, and the winter of hibernation, cannot be denied. Perhaps, indeed, a relationship between the weather and the arts might be possible.

"An obvious pattern exists in the seasons themselves . . . the creative arts . . . form is *only* a pattern composed of rhythmical designs. If the weather affects the arts, the composition should clearly show this influence.

"The cycles of climatic changes have been charted by Wheeler; the divisions in artistic development have been outlined by me," Dr. Olmstead went on. "When the two graphs, one of weather, the other of art, are compared, they are startlingly similar. Especially in short periods of about a decade, and occurring at about the same date, they both seem to change consistently; the weather, first wet, becomes dry at the same time that the artistic composition once open becomes closed."

Ever since man began looking at nature for inspiration, art has been related to the changing climatic cycles. When the weather was predominantly warm-wet, landscapes were filled with the blossoming flowers of spring. When it was warm-dry, the ripened fruit of summer was the subject. When it was cold-wet, the broken branches of autumn appeared. And when it was cold-dry, the period when man defies the weather most of all to rely upon himself, art emphasizes *man's own handicraft.* Nature, as the source of inspiration for art, is comparatively recent in European history, but from 550 B.C. onward, art has always followed these trends according to the climatic cycles. These historical facts compose a new theory of art.

General Characteristics of Dry Art

massive
horizontal
three-dimensional
smooth
closed
static

In dry times the sky is clear; the sunlight is brilliant. Nothing moves; nothing changes; all is calm. At midday there are highlights; at midnight there are low-darks. The eye is capable of clearly distinguishing each object in its entirety, along with its mass, its surface, its outline—all definite and distinct. The close-up looms large, while the distance is clear.

Dry art should be closed and static, simplified and calm. Like the cloudless sky, its surfaces should be smooth, its main direction horizontal, resembling the horizon. Clearly defined by the light, it should have contrasting values, definite outlines, clear and massive volume. Typically dry, its chosen colors should be orange and red. With an intellectual subject, this dry art should be rationally composed by objective artists.

General Characteristics of Wet Art

fragile
vertical
two-dimensional
uneven
open
dynamic

In wet times clouds fill the sky, cover the sun, and reduce the light. This atmosphere produces a diffused monotone. When the rain falls, it blurs the objects and merges all nature into a general haze.

In this wet period the artistic composition should, it would seem, be open, dynamic, and tempestuous like the winds. Like the rain, it should be vertical. Because of the clouds, it should be monotoned, with indefinite outlines, shallow space, and fragile volume. Like water itself, the primary color should be blue-green. Its subject should be spiritual, rapidly executed by a subjective artist with an emotional attitude. The favorite medium should be water; it should flow upon paper.

General Characteristics of Warm Art

solid
curvilinear
projecting
thin
expansive
surface

In a warm land, minus personal worry over material necessities, man can discard his garments and reveal his external body. Strong light rays project the objects around him and tend to whiten everything with a glare.

A hot art thus should have a superficial style with an expansive composition and an opulent pattern. Golden yellow, like the heat, should be the favorite color; and because of the strong sun, the lights should be crowded. Surfaces should be thin, like the skin of men.

The lines should be curvilinear, resembling human plumpness. Their direction should be external, with the same flexible outline which the artist sees in his own body. Space should be filled with a projecting solid.

General Characteristics of Cold Art

hollow
angular
receding
thick
contracted
detail

In a cold land the protection and preservation of himself and his family would be the primary necessity which each man must maintain. Erecting shelters, building fires, making clothes, finding food, would keep him busy. If he is successful in this, he can enjoy his security, but he has himself and his work to thank for his well-being. With less sun, dark shadows would predominate, especially within the corners of his home.

A cold art thus should contrast with a profound composition and a frugal pattern. Violet, like the shadows, should color the world. The concentration should be on the darks, while the surfaces should be thick, like fur. The lines should be angular, resembling the home man has constructed. Space should recede, and volume should be hollow.

With a subject devoted to artificial, man-made products, this cold art should be personal and should be composed by individualistic artists who demand credit for their work. Oil is a possible medium, while the technique should be related to building with wood.

The pure wet style attempts the impossibility of becoming two-dimensional, especially in temporary and open structures. Walls, when they appear, are dynamically uneven. On the other hand, the dry style is perfectly three-dimensional, static, and massive; here are the permanent buildings, correctly closed. The warm attitude—impersonal, idealistic, communal—has thin walls, so that the structure seems to expand outward. The cold style, however, better fulfills the primary architectural functions, with its thick walls, contracted space, and emphasis on the interior; this type is particularly secular and domestic in intention.

As we have seen, Wheeler identified warm-wet phases with times when the organismic culture pattern, emphasizing wholes as opposed to parts, reaches a climax:

Literature: tragedy, philosophical poetry, time and fate emphasized, subordination of the individual to higher forces, mysticism, "classicism," social novels, reflection of socialistic trend in society, free and blank verse, abstract movement.

Costume: regimented, severe, simple, streamlined, regal, reserved; conservative use of line and color.

Painting: regimented use of color, monochrome, abstract movement, symbolism, subjectivism, impressionism, caricature, distortion, fantasies, surrealism, cubism. When the weather was warm-wet, slender flowery forms resembled those of spring, whether artists copied nature or not.

But how do these generalizations tally with the more "permanent" arts?

Architecture has always been more traditionally limited by the past than the other arts. Often older forms evolved by peoples in their earlier homelands have been retained, even after they have invaded new countries. There is also a time lag in architecture behind that of the other more quickly executed arts; because in building, the original plan and the finished product are often widely separated in date. A great cathedral takes several generations to complete, but even the most grandiose scheme in sculpture, unless architectural, belongs to only one man's lifetime. In addition, architecture is rooted in the land and retains its visual importance there until purposely destroyed. Buildings remain long after all else is destroyed by time; foundation is often all that remains for the archaeologist. Finally, conservatism has always been more prevalent in architectural designs, because additions and remodelings are acceptable here, whereas restoring and retouching are disallowed in painting and sculpture. For the above reasons, architectural periods are longer in time, with fewer major changes than in the smaller arts.

But climatic changes occur comparatively frequently. Though a century might be predominantly of one temperature or precipitation, there are also interruptions of other types. In other words, some forms of the architectural style have their origin in a climatic time or place far distant from the date or locale of the actual building. Other forms out of the immediate or far past are retained even though the weather period has in the meantime changed. But the evolution of an architectural period should in some respects correspond to the contemporary climatic conditions. The basic forms already noted for painting are thus to be found in architecture, but in modified form.

Amazingly enough, warm-wet periods have produced an architecture which in the history of art has been greatly admired, especially in our own recent warm-wet times—perhaps for the very reason that its existence is a contradiction in terms. Neither as cold art, for man's protection; nor as a dry form, which is enclosed; but rather as an open warm art and as a wet impermanent form, this is the architecture of hopeful imagination.

The reed hut is a typical warm-wet architecture which seems to be the basis of some early building in Mesopotamia and Egypt. In the swamps of the rivers, tall thin stalks grow up into the sky from the water below. When their tips are fastened together, a fragile covering is formed, in shape like a pointed arch. When the stalks themselves are tied together, columns are made. With such willowy materials it is impossible to construct straight lines and right angles (the usual characteristics of commercial architecture today), so that the ground plan curves as much as does the elevation. It is possible that these earliest reed huts were drawn together over some slight round hillock of drier land, rising above the marshes, so that the walls were thus anchored in the water, but the floor was the somewhat solid ground.

Warm-wet architecture with its upward-flowing movement and curvilinear outline visually resembles rolling hills. Some such geographical locale should be the homeland for European warm-wet architecture, where rain and soil together produce a growth of vegetation in fertile fields and valleys. Certainly warm-wet architecture was often given such a setting to further emphasize its own rounded and upward movement. This same form is repeated in the elliptical ground plan. The column, imitating the growing stem of a plant or tree, is the support for the structure; and like the entwining upper branches of shrubs or trees is the pointed arch which results in a vaulted roof. With its emphasis on colored variations, and because painting is the primary warm-wet style, murals can also be employed.

Obviously, the early Gothic is an excellent example of the warm-wet style, where the pointed arch and the ribbed vault rhythmically move upward to a heaven above. The ground plan is almost elliptical, what with the greater emphasis on the apse and the opening of the transept. Stained-glass windows pour soft light into an interior which is painted wherever possible. Piers, inherited from the Romanesque, lose their weight, and are adorned with slender half-columns which move upward. High vaults emphasize this growth; and on the elaborated exterior, soaring towers touch the skies. Buttresses fly into space. Everywhere air and light dance through the form in complete disregard for functionalism.

In some respects the Rococo resembles the Gothic. Eighteenth-century buildings are often light and airy with large windows. This interest in open spaces caused the creation of many large squares, such as the Place de la Concorde in Paris. In general the structures are smaller and more intimate than those of the preceding century. The decoration, charmingly graceful, freely irregular, is based upon the warm-wet curve.

For the early twentieth century, the American skyscraper is the best example of warm-wet architecture. Less functional than is generally supposed when the many rooms cannot be rented, these buildings soar high not only because of the invention of the elevator, but also because man wanted to look upward. Walls have been opened, because the warm-wet attitude demanded this novelty. As a result, new techniques of steel frame and the cantilever system were invented. Landscape gardening, so popular recently, is thus only a reflection of the usual warm-wet interest in natural growth. Other characteristics of warm-wet architecture: streamlining, functionalism, harmony of medium and design, harmony of structure and environment, stylism.

Warm-wet sculpture has quite different intentions. It embodies impressionism, distortion, abstract studies of form, rhythm, masses; unification of form and medium; idealism, gods, abstract themes. Primary is its sympathy with the spring of growth. Thus earth itself—in more technical terminology, clay—becomes the material for warm-wet sculpture. Mixed with water, clay can be modeled with a light touch. Interestingly enough, vegetable life is also more two-dimensional in contrast to the warm-dry three-dimensional subjects of animal and human life. Or if the subject is human, elongated ladies dance lightly with billowy drapery. In this same elliptical form, arms are flung upward into the air, and feet are precariously balanced on tiptoe. Moving in space is the result; consequently, holes are cut into the material through which the winds are allowed to play.

Because painting is the primary art of the warm-wet period, sculpture is often tinted and gilded. The sculptor no longer prefers highly polished surfaces, but so lightly finishes his figures that they remain highly suggestive.

Obviously Gothic statues best express this warm-wet style, especially the Madonna and Child, so popular at this date. Her swaying body, rippling drapery, tilted head, smiling lips, all give her a lighthearted joy in momentary movement. Her fingers open; her jeweled crown is decorative; her arms reach out from the body; and her figure moves away from the former cold-dry imprisonment against the wall. Other figures of this period have the same long, slender neck; delicate limbs; the tall, willowy forms; all of which make this type particularly fragile.

For a short period in the nineteenth century, the Impressionist Auguste Rodin returned to this warm-wet style, employing clay so soft that the surface ripples with fingerprints. Again entwined figures and opened limbs correspond to the new dynamism.

In the twentieth century the similar clay type of elongated and elegant figure is repeated by Gaston Lachaise, and the curvilinear movement by Constantin Brancuşi. Most interesting perhaps is the quite different style by Henry Moore, in which holes have been opened through the figures themselves to allow the extreme of warm-wet movement through space.

11

Warm-Dry, Cold-Wet, and Cold-Dry Arts

Warm-dry architecture, which rivals sculpture, has had a great vogue among connoisseurs, probably because its idealism negates practicality. Most traditionally "great" monuments are warm-dry in concept and date, regardless of the longer stylistic period to which they belong.

Only in a low plain, where the sun is hot, the rains infrequent, and with no interfering hills, can a building truly dominate the landscape. The round arch, so characteristic of this form, was apparently developed in Mesopotamia. Here the smooth surface can best glisten in the sunlight, aided by shining mosaics. In the perfect warm-dry type, a circular ground plan carries a round dome, which actually does compete successfully with sculpture because of its larger, grander scale.

Hagia Sophia (its very name, "Holy Knowledge," generalized into the typical warm-dry abstraction) is an excellent example. This most famous domed structure in world architecture was built in the reign of Justinian, a warm-dry dictator. Later Muslim mosques have gleaming mosaic-covered domes, which contrast with the low barren landscape and appear as jeweled masses from the far distance.

A short warm-dry phase, from before 790 and from 800 to 810, produced Charlemagne's chapel at Aachen, where, today at least, the building is the dome. Parts of the eleventh-century Romanesque were warm-dry, and the large ground arch reappeared with an occasional dome. In the warm-dry mid-sixteenth century the dome became supreme, particularly in St. Peter's at Rome. Subsequently, any warm-dry phase resulted in the reappearance of the dome.

When it was warm-dry, the compositional elements were round and firm, though fruits might never be represented. To best fulfill the warm-dry ideals, sculpture has usually been made of marble, since that is the most solid and permanent of all earth materials. Also, rocks are often rounded and even polished by glacial or marine abrasion, and thus in their natural state suggest the final form of the statue. Sometimes only a slight manipulation will transform the stone into sculpture; and even when roughly quarried in more civilized periods, the artist retains this original idea by posing the arms and legs to enclose the body, bowing the head, bending the knees and elbows, and adding a final, finished polish of bright gleam.

The usual admonition that a "statue must be able to roll down a hill" is the warm-dry ideal. Because extraneous elements, such as clothes, would negate this ideal, and because the demanded "pure" art of a warm-dry attitude rejects all superfluities, the nude without man-made additions is the usual subject. In the history of art, nudes are more common in the warmer countries; that is, Greece had more than Rome; Italy and France have more than Holland, Germany, and England. How rare are nude statues in the colder United States! Indeed the "perfect" figure is even further limited to the female, which is the roundest of all possible forms in the imagination of the sculptor. Again, the farther south, the more popular the female; the farther north, the more predominant the male figure.

Obviously the classical Greeks best fulfilled these warm-dry desiderata for statuary, and hence their supremacy has never been questioned (although art historians must admit that in most periods of Egyptian and Babylonian sculpture, the forms are even rounder, fuller, firmer, and have a higher polish). In the European tradition the Grand Style (what a warm-dry term) of Michelangelo best exemplifies these characteristics, and his preeminence is taken for granted—although at the beginning and end of his life in temporary warm-wet and cold-wet phases he produced a different type of art which is less generally known and recognized. He was followed by the mannerist Giovanni da Bologna, technically superior, idealistically inferior.

The warm-dry tradition of statuary was almost lost during long periods of either wet or cold art, until it was revived with all of its characteristics by the neo-classicist Antonio Canova at the beginning of the nineteenth century. Here again are all the features of the warm-dry style—polished marble, the feminine nude, round form, idealistic concept. Subsequently, during the long cold of the middle of the century, such statuary disappeared until Aristide Maillol reintroduced these characteristics at the warm-dry end of the century.

The twentieth century, in spite of its long warm-dry period in the thirties, had in the intervening years reacted against these academic doctrines (the cold destroying the warm), so that the great sculptors seldom completely exemplify the pure warm-dry statuary in all its respects. Yet Ernst Barlach emphasized mass, and Pablo Picasso the abstract roundness.

But because of the warm-dry insistence upon the eternal qualities of full, round sculpture, an unfortunate situation has occurred in the past century and a half. Warm-dry academic standards have adversely affected creative painters, but have not completely stifled the art. Yet sculpture, until the recent warm-dry phases, was almost entirely destroyed because of the refusal to accept any style which did not fulfill the basic standards of the highest warm-dry ideals, no matter how cold or how wet the weather conditions. This is particularly true of the United States, but even in Europe there is a great lacuna between the warm-dry, neo-classicism of Canova and Maillol, only partially filled by Rodin, who worked much of his life during a warm-wet period. Early twentieth-century sculpture, which revived in the warm-dry period just past, has thus often been a strange agglomeration of many styles.

The Cold-Wet Phase

Raymond Wheeler observed that in cold-wet phases, people change from time-mindedness to space-mindedness. Human moods change from austerity, pessimism, and lethargy to gaiety, optimism, and assertiveness; a "classical" changes to a "Romantic" period.

There is a shift of cultural output from the organismic to the mechanistic pattern across the entire realm of human activity:

Literature: lyric poetry; simple, rhyming verse; descriptive novels about simple and common people in their everyday life; importance of the free individual and the expression of his feelings; assertiveness portrayed; virtue of hard work and individual enterprise extolled; revolutionary in the sense of demanding democracy; love frankly expressed; simple in meaning; sentimental rather than philosophical; forward rather than backward look.

Costume: Elizabethan, Victorian; elaborate in use of color and design; frills, laces, scallops, bustles, hoop skirts, ribbons; greater modesty than in warm times.

Painting: concreteness; naturalism; imitation of nature; realism; humble subjects (animals, birds, flowers, landscapes); simple love themes idealized; emphasis on elaborateness and accuracy of detail; the part stands out more in its own right—less dependence upon the whole for meaning. When it was cold-wet, the forms disintegrated into diagonals, as though the entire composition were blown and broken by the wind.

Architecture: elaborate; baroque; highly adorned; rococo; garish color and design, "gingerbread style"; reaction against the regimentation, simplicity, and streamlining of the warm phase "classical" styles; often lack of unity from overemphasis on the part.

Historical examples of typical cold-wet architecture are rarer than those of the warm type, and are less easily differentiated from the cold-dry style than the warm-wet were from the warm-dry. The major form is a simple tent—comparatively good protection, but so easily destroyed that it is not readily discovered by archaeologists.

There are many other explanations for the rarity of a cold-dry period. Destructive in intent, cold-wet periods seldom sanction the labor necessary for major architecture. Fundamentally impermanent in attitude, there has been little justification for maintaining the comparatively few structures which were built. Yet elements from out of a cold-wet past remain as the basis for most European architecture—pitched roof, gabled facade, and wooden material.

Cold-wet architecture belongs to mountainous regions where the frequent storms are frighteningly close, where the slashing rains produce torrents down the steep slopes. To protect the buildings, this same downward movement is used for the gabled, pitched roof, and the resulting pyramid volume (if the building is more three-dimensionally dry than two-dimensionally wet in style). In order to cling to the sides of the mountains, the form of the peaks is often repeated in a triangular ground plan. Mountainous countries such as Switzerland and Scandinavia still have the most characteristic examples.

The early Christian basilica has many of the cold-wet forms, although combined with cold-dry elements. Particularly is there an emphasis upon the interior, which, though rectangular, is so long and narrow that it appears almost triangular in perspective. The gabled roof for the center, with its rafters breaking the surface, reappears cut in half for the side aisles, so that sharp angles are everywhere. The ground plan, too, is broken into long narrow sections—the nave, the aisles, the bema—though set at right angles in the more usual cold-dry type. The pier, too, would have been more common if Roman columns had not been available. Here is another instance of the shoddy work of the cold-wet attitude, which reused whatever was at hand and put all together in a flimsy manner.

The Romanesque, in many respects, was a cold-wet architecture, particularly in Germany. Although there were obvious inheritances from the cold-dry earlier Middle Ages and although the revival of warm-dry forms occurred in the eleventh century, the general impression is one of angular irregularity, better exemplified in the sculptured reliefs attached to the walls. Even the groin vault has a downward thrust, although combined with the past warm-dry round arches. Again, little remains untouched by later additions; and few of the original plans are known—or accurately datable.

So, too, did the Baroque destroy the Renaissance, although there are inheritances from the individualistic cold-dry fifteenth century and the authoritarian warm-dry sixteenth century. "Broken" is the most descriptive word to be applied to this architectural style—broken entablature, broken pediment, broken moldings. These breaks, so typically cold-wet, like the recessed surfaces, cast dark shadows upon facades and within interiors, so that Baroque buildings are excessively somber. Everywhere there is angular movement. Climatically, however, the seventeenth century in Italy, where the style originated, is warmer rather than colder in style, so that there are many curved forms beside the angular. This is also the period of steps, built up and down the hills of Rome.

Whenever the weather became cold-wet, there was a revival of medieval forms, such as occurred in 1510 and 1810. Already there are new tendencies away from the warm styles of the past fifty years toward the cold styles of the present and future. Probably the most interesting of the contemporary cold structures are those of the Buckminster Fuller Foundation. As outlined in the lectures of their fellow, James W. Fitzgibbon, their research began with spheres (obviously still warm), then proceeded to struts arranged assymmetrically, and finally have resulted in an amazing new balance achieved by many small props, structurally so sound that a new architecture could be the result.

Sculpture: real people instead of gods; children, domestic themes, gaiety, realism, naturalism, play, song, concrete subjects; absence of distortion.

In cold-wet sculpture the artist, armed with a sharp instrument such as a chisel, hacks out his design. Destruction, not construction, separates cold-wet sculpture from the opposite extreme of pure warm-dry statuary. Death, not birth, distinguishes the cold-wet from the warm-wet attitude; descending, not ascending, is the movement. Angles replace curves, and there is deep cutting instead of projecting modeling. Most typical of the cold-wet period are the graphic arts; and similarly, the most extreme sculptured form is relief in intaglio.

Metal is probably the most employed technique because, unlike warm-wet clay, warm-dry stone, and cold-dry wood, it is artificial. Sharp edges, angular lines, arranged in triangular form, help describe a subject matter of male destruction.

Since there have been few long cold-wet periods—and since the cold-wet phase is least sympathetic to sculpture—there are comparatively few well-known examples. In the Romanesque period the popular *Dead Christ on the Cross*, terrifying in its inhumanity and horrifying in its destructiveness, is a good illustration. Set against the cross, with drooping head, outstretched hands nailed to the beam, and feet fastened at the base of the long thin board, the form is predominantly triangular.

For the seventeenth-century Baroque, Giovanni Bernini's *St. Theresa in Ecstasy* is cold-wet in both style and date (1646). Downward comes the sharp shaft and pointed arrow to pierce the heart of the saint as she swoons away. Intense, even destructive, is the subject matter. The limbs are angular; the drapery moves; the outline is broken.

Revivals of medievalism and returns of expressionism have employed this same type, as in the teens of the sixteenth, the nineteenth, and the twentieth centuries. In the last-mentioned cold-wet period, Picasso invented Cubism and sculptured a head so cut out in form, so triangular in shape, so depressing in concept that this style has not yet been accepted by the mass public. In so-called modern sculpture, rough surfaces, angular lines, and depressed composition have appeared in metal sculpture of wires and pipes.

The Cold-Dry Phase

Architecture, by its very reason for existence—living space for man inside walls which protect him from the inclement weather outside—is the major cold-dry art. Its original invention undoubtedly occurred in a long, cold glacial period when protection was indeed necessary. Architecture first gave man his sense of power over the elements; similarly the crude early graves are also a defense, even though superficial, against the coldness of death itself. Pride in such ingenuity to erect his own dwelling place, rather than in utilizing the natural caves, resulted from man's practical application of knowledge learned from experience and necessity. Each man needed this protection as much as he needed his own clothes; and so in time, a community of houses developed into a village.

Cold-dry architecture belongs to high elevations where thick walls are necessary against the low temperatures, and flat roofs are no inconvenience because of the scant precipitation. The typical rectangular form, with the emphasis on the interior, also belongs to this locale, where surrounding mountains are natural walls against the occasional winds and rain, and the plateau within is literally well enclosed. Thus the ground plan is square, and the interior has a cubical volume obtained often by the post-and-lintel system of support. As another protection against the cold, secondary layers are added as insulation so that the surfaces are textured. Tapestries hang on the walls; rugs are spread on the floors. Decoration added to these buildings must follow the primary horizontals and verticals. Space is divided into many sections or rooms to further separate the exterior cold from interior heat.

The cold-dry monasteries of the late Dark Ages were fortresses which belonged to the church, but also protected a large populace. The plan is approximately square, cut by other rectangular units placed close together at the front, back, or sides of the central church to form a community. This was also the gridiron plan of the old Roman military camp, and its practicality is obvious with crisscrossing streets for easy communication within the entire area. Similar fortress-castles were designed for the nobility and their dependents.

Fortresses also were the fifteenth-century Renaissance palaces, with forbidding exterior, thick walls, and the rustication of the textured surfaces. Shops below had a practical purpose, and even the overhanging cornice is protective. Visually the flat roof was emphasized by long, horizontal string courses. As a result the interior was pleasantly cozy, though dark. Interestingly enough, in the central court which actually provided the light, the slenderer, thinner, curvilinear forms of the warm-wet style were employed to correspond with growing plants. As would be expected in every cold-dry period devoted to business, these palaces belonged not to rulers, but to great financiers like the Medici.

Although architecture is the most typical of the cold-dry arts, interestingly enough there have been, in our own contemporary ideology, few so-called great periods in the last thousand years of this art. The longest, coldest, driest periods mark the division between major eras five hundred years apart; these are found in the tenth century, the fifteenth, and have again occurred in the latter half of the twentieth century.

Occasionally, cold-dry forms have reappeared. The interruption of warm-dry Mannerism in the sixteenth century produced Pierre Lescot's facade of the Louvre; and the short cold-dry phase of the warm-wet and cold-wet Baroque saw Claude Perrault's facade. In mid-eighteenth century England the simple Georgian temporarily replaced the warm-wet Regency. The ideals of warm-dry (i.e., "divine art") have so influenced our standards that the mere practicality of cold-dry architecture hasn't been enough to justify their existence. There are, however, thousands of minor examples, particularly in domestic architecture. Unfortunately, in the United States the original functionalism of the earliest settlers was discarded as soon as the colonies became a country and its inhabitants became wealthy. American architecture then became imitative—classical temple-mansions, romantic cathedral-houses, Romanesque fortress-dwellings, Renaissance palaces, Baroque villas. Probably this has been caused by the past fifty years of warm weather, which naturally have reflected other similarly warm periods of the past.

The future has become cold-dry again. The cold-dry period should be preeminently our own; and indeed, the rectangular "box house" is still the choice of most of our middle class. With the situation so confused, the future would appear dark, except that a few great architects such as H. H. Richardson, Louis Sullivan, and Frank Lloyd Wright, in spite of the warm weather of the recent past, still retained many basic cold-dry characteristics.

Cold-dry sculpture, as already noted above, is primarily a rectangular relief with enclosed background attached to architecture. Like the hammer, which the builder uses, the mallet will be one of the major tools. Wood is probably the most typical material, since boards in buildings can be easily carved out, as decoration. When it was cold-dry, pieces were put together again by man into his own handicraft.

There are many possible subjects, but especially ordinary people will be portrayed as portrait heads in a realistic style because of the characteristic pride of man in himself in cold-dry times.

Relief sculpture, however, because it has a flat background, is basically a decorated wall, and thus is related to architecture, which dominates a cold-dry period. It is like the second layer of a tapestry applied to the structure, but it is less a protective covering against the cold and more a proof before man's eyes of his own ingenuity and handicaft. Thus, unlike a statue, a relief exists only as long as the walls of the building stand; and one broken part can ruin the meaning of the entire piece. Consequently, for relief sculpture any type of subject matter is usually allowable, unlike the limitation to human and animal life for statues. Because it is not three-dimensional, line is more important than mass, and thus reliefs also occur in wet periods. If relief is employed, everyday work of men by which they survive the intemperate climate will be shown. Sculptors themselves in cold-dry periods are more apt to follow a related practical profession and become craftsmen or designers, such as goldsmiths, silversmiths, and woodworkers.

The major cold-dry period in the past was the fifteenth-century Renaissance, but the Roman background was also cold-dry. Lorenzo Ghiberti began it with his bronze doors, although the actual style varied with the changing weather. Andrea del Verrocchio employed it with his reliefs, his realistic portraiture, and his particular interest in the minor arts. Benvenuto Cellini continued it a hundred years later, growing more famous for his saltcellar than his statues. In the United States Paul Revere is remembered not only for his patriotism in the cold-dry period of the American Revolution, but also for his silverwork.

This cold-dry type should be the sculpture of the future; but it is to be wondered if it will reappear, because the warm-dry ideals for statuary have been so thoroughly accepted. It is true, however, that in the United States the cold-dry eighteenth century produced the practical William Rush, who designed figureheads for ships as well as realistic portraiture. But for the mid-nineteenth century there was no sculptor whom we recognize as great, although John Rogers, first an engineer, then a machinist, eventually became a realistic sculptor whose figures were popular in the Victorian era.

Realistic sculpture is considered the only possible form by the uninformed public. It is probable that for some time we will retain our two opposing schools of sculpture, just as the Romans did: one, a warm-dry imitation of the great classical standards of Greece and France; the other, a cold-dry realism which is ours by tradition and climate. In the truly practical minor arts, however, where the Romans also excelled, and which our machine mass production has emphasized, there is already so much progress that probably here the finest new art will be created by future designers.

12

The Cyclic History of Music

The evidence gathered since Wheeler's climate-and-history project was started in 1932 now helps to delineate five great periods of music history, each defined by cataclysmic changes in weather and culture. Each music period in a 500-year cycle, since the beginning of Western thought among the ancient Greeks, comprises within it the five 100-year weather cycles which show in each century four unvarying rhythms in this order: 1) cold-dry, 2) warm-wet, 3) warm-dry, 4) cold-wet.

As earlier pointed out, with each shift in the cycle come corres-

*Adapted from Warren Dwight Allen, "Music History in Five Five-Hundred-Year Cycles (Greek-Early Christian-Gregorian-Medieval-Modern —Conditioned by the Parallel Climatic-Cultural Cycles)," *Journal of Human Ecology*, Vol. I, No. 13 (1951).

ponding changes toward opposite poles of political control and of artistic expression. In warm times formal arts express the *ethos* of the whole society under varying degrees of socialistic, collectivistic, sometimes tyrannical forms of government. Thus warm-wet music is characterized by impressionism, intellectualization, "spiritualization," dissonance, atonality, individual tones lost in total effect; abstractness of theme, formality; orchestras and bands; regimentation of musical composition, symphony, chamber music.) In cold periods democratic tendencies assert themselves; romantic *pathos* of expression is then more concerned with the spontaneous feelings of individuals. (Cold-wet music displays comic as opposed to serious opera; sentimentalism, melodic music; easy to understand, concrete, emotional, rhythmic (simple rhythms); baroque, adorned, bombastic, show-off, interest in individual performers as opposed to bands and orchestras; program and descriptive music; each note composed to be heard, emphasis on clarity and precision with reference to the part.)

These terms are perilously flexible, but convenient, as Curt Sachs points out in his *Commonwealth of Art*. It may be more exact to say that in warm times the whole *tends* to determine the parts; that in cold times the parts *tend* to determine the whole, in art and in society. So, for convenience, throughout this summary the word "classic" will be used with reference to organic forms in which the whole seems to dominate the parts (the word "classical" is widely associated, rightly or wrongly, with a late eighteenth-century period). The adjective "romantic," as generally accepted, is bound up with the personal-emotional and will be used with reference to music in which the parts in succession tend to make up the whole (as in the decorative phases of Baroque, Rococo, and so-called gingerbread architecture).

The five great 500-year cycles, each beginning and ending in cold-dry years, with the emergence of definitely new musical styles and habits, are *approximately* as follows:

1. 576 B.C.-A.D. 30. From the music of Athenian drama to the Christian revolution.

2. A.D. 30-460. From the syntheses of Christian and pagan

melody to the fall of the Roman Empire with the control of music taken over by the authoritarian church; with Oriental chant substituted for the Greek styles degraded by Rome.

3. 460-960. From the Gregorian era of official, quasi-Oriental plainsong to the revolt against monody and the beginnings of European harmony (in music only, of course).

4. 960-1475. From the Romanesque and Gothic eras to the Renaissance, with the gradual restoration of musical freedoms sacrificed during the Dark Ages.

5. 1475-1980(?). From the beginnings of music as we know, use, and hear it to the present and immediate future; from the establishment of our modern tonal system to modernistic revolts against it.

Music does affect, and is affected by, political, social, economic, artistic, linguistic, philosophical, religious, and ideological changes. Wheeler's epochal discovery is that all of these are affected together by human ecology.

Wheeler's evidences of cyclic cultural-climatic change must first be understood on the macroscopic scale, hence the limitation of this first application to music history to the five great 500-year cycles. It is not necessary to prove, for example, that J. S. Bach achieved the classic, organic quality of his Mass in B minor by writing it during hot weather and the romantic, highly personal qualities of his chorale-preludes by playing the organ on cold, rainy days. The important facts for history are 1) that his genius was particularly at home in the ancient democratic folk tradition of the passion play, during which the congregation sang folk hymns or chorales, which became official Lutheran music during the cold period of the Reformation; and 2) that he was able to furnish proofs of his vast versatility, even in the classic tradition, by turning to the organic form of the Roman Catholic mass, for an autocratic sovereign who was probably at his worst during hot summers in Saxony!

We must also be careful to define our terms so that there is not too much conflict with those long in use. But there will have to be considerable study of the terms "classic" and "romantic" as they have been applied to periods. They belong not merely to periods, but to style; and our whole system of "style periods" is due for reexamination. For example, there is a vast difference between the classic integrity of Bach's "Sanctus" and the integral wholeness of a piece in the modern twelve-tone technique. Each is a whole greater than the sum of its parts, but—what a difference! Similarly there is a cosmic gulf between the romantic "Crucifixus" in the same Bach mass and a piece of boogie-woogie. The latter is a ground bass with a hangover; the Bach masterpiece also has a bass part constantly repeated; but what a difference in value, use, and function!

On the macroscopic scale the climatic-cultural cycles disrupt none of the traditional divisions of history; they reinforce and clarify our concepts of ancient, medieval, and modern. The phenomena of Renaissance and Reformation can also be more clearly envisaged, also the changing aspects of the important Baroque era. The eternal swings from what we have called the classic and the romantic have only been touched upon here. Thus an intensive study of Wheeler's 100-year cycle may be helpful in the near future.

At any rate, my own excursions into music history have convinced me that in the principles of human ecology we do have, in fact, the basis for a much-improved understanding of great musical movements and trends.

In our own day, some composers have been extremely sensitive to cyclic changes. Igor Stravinsky, notably in his return to neoclassicism after 1920, reflected the warm trend. Now it will be interesting to see how and whether his older, postromantic tendencies return in line with Wheeler's predictions. Henry Cowell's super-classic tone clusters, in which only the top melody stood out, were abandoned some time ago. Peculiarly sensitive to new trends, his inspiration now comes from the riches of romantic American pioneer music.

American music education in the nineteenth century was largely devoted to the training of soloists; during the warm 1930s it was devoted largely to the socializing of music in massed bands, choruses, and orchestras. Until recently the orchestrations of Tin Pan Alley have obscured melody with rich, lush arrangements; now the hillbilly and his "gittar" are again in the saddle.

Art may have to be intelligible once more. Techniques have improved so greatly that all sorts of communications, no matter how confused, can be conveyed. Musicians can play anything set before them, whether it is worth playing or not. One of the greatest of our conductors says privately that he has played certain modern works *without understanding them.*

One reservation must be made: History must not be content with recording the phenomena of change. The history of music now in preparation will deal also with the stable *continuities* of value, use, and function. *All* of the origin theories advanced in romantic histories were, and are, valid—each for a particular style and use of music. Throughout the cycles, men have continued to make use of certain basic musical idioms—for lyric song and dance, for fighting and commanding, for praising and lamenting, for the display of skill in virtuosity, in the chordal harmonies of simultaneous effort, and in the cooperative counterpoint of different parts. The Greeks initiated many of these that are present in varying degree in every cycle except that containing the Dark Ages. *Romantic* styles always define very clearly the basic continuity used in a piece of music; hence their popularity.

It will be a great day, even if cold, when clearer communications are in order—in *all* arts, music and politics included. These clearer communications will not be repetitions of old simplicities. The music of the future will gain in depth and power from the bitter experiences of discordant extremism; some of the work of our younger composers is already doing so.

13

The 500-Year Cycle and Climate-History Ratios

"*There seems little doubt that for the last several centuries long cycles of* about 1,000 to 1,020 and 500 and 510 years have been important. A number of scholars have observed a culture cycle of 500 years," John Cejka told me.

As a result of his extensive research, Dr. Raymond Wheeler was able to chart 500-year climatic cycles throughout history. Wheeler found that, both climatically and culturally, secondary climaxes of cold temperatures and devasting droughts, together with great waves of migrations and drastic revolutions in society, have occurred at the end of each intervening 500-year period.

Every fifth cold phase is an unusally severe one, as demonstrated by the fifth century B.C. and the first, fifth, tenth, and fifteenth centuries A.D. Wheeler's research indicated that the sixth century B.C., the fifth century A.D., and the fifteenth century A.D. were also centered on unusually cold and dry times. They were centuries of widespread migrations; periods of low lake levels and slow tree growth. All three were characterized by the same process—the death of an old world and the birth of a new one.

The main divisions of history—ancient, medieval, and modern—serve also as landmarks in the history of world climate, Wheeler said. "Old civilizations collapse and new civilizations are born on tides of climatic change. The turning points occur when cold-dry times reach their maximum severity."

Every other termination of the 500-year cycle has coincided with the termination of a 1,000-year one. In the center of each 1,000-year cycle there seems to be a much warmer period. The first such period in our historical epoch reached a climax at the time of Julius and Augustus Caesar, about 40 B.C.; the second, around A.D. 900 and 1000. During the second of these warmer periods, Greenland and Iceland were colonized. Once the warm phase of the long cycle was on its way, Greenland froze up and has remained so ever since. Many times in the fourteenth and fifteenth centuries, the Baltic Sea froze over between Germany and Sweden, permitting horses, even armies, to cross from one country to another on the ice. Every alternate 500-year cycle has ended in the middle of the 1,000-year rhythm during its warm phase. However, there has always been a complication: The 500-year cycle breaks down into shorter periods that last about 100 years on the average, as we have seen. These clearly defined cycles each have a major warm phase and a major cold phase, two major wet and two major dry periods. (One major wet period occurs at the beginning of the warm phase, the other at the beginning of the cold phase. One major dry period occurs at the end of the warm phase, the other at the end of the cold phase.)

Raymond Wheeler believed that each great revolution of history, synchronous with the termination of a 500-year cycle, has been characterized by a great advance in democratic benefits to the common people.

The year 575 B.C. marked the termination of both a 500-year and a 1,000-year cycle. Up to that time, the inhabitants of Europe were largely still Stone Age nomads. Little enlightened thought had yet existed anywhere in the world, with the exception of rudimentary engineering, metallurgy, and agriculture in Egypt. China and the older civilizations of Mesopotamia had fallen into decline. The last strong power to assert itself had been Assyria, and its final brilliance had already run its course. The older of the ancient empires, with their god-kings, were gone forever, together with the great tombs, temples, and religions which had sustained them.

The beginning of the mental attitude and intelligence which would produce modern civilization began in 575 B.C. Primitive techniques of structuring mental processes gave way to such Greek philosophers and scientists as Thales and Pythagoras, who prepared the intellectual climate for Plato, Aristotle, Euclid, and Archimedes.

This was the beginning of scientific knowledge; the beginning of a rational search for truth; the beginning of mathemetics and logic; the beginning of serious analyses of social, political, and ethical problems; and it was the beginning of democracy as we know it. There was also an industrial revolution, the emergence of new crafts, enlarged perimeters of trade areas.

During this 500-year cycle, the centers of civilization shifted from south and east of the Mediterranean to the peninsulas on the north. This period was dominated by Greece, Macedonia, and Rome, with Persia steadily offering competition. The Greek city-states, such as Athens and Sparta, had their Golden Ages. The cycle of 575 B.C. to A.D. 1 saw the growth of Greek democracy —emerging from the convulsions of the sixth century B.C.

As the city-states declined, Rome gained in power. Rome's greatest contribution came from her genius for law and organization, but she had reached the climax of her power just before the 500-year cycle drew to a close.

The next 500-year cycle of our historical epoch ended at the time Jesus Christ was born. Until A.D. 70, much of the first century was cold. The great Roman Empire, which had taken five hundred years to build, had passed her cultural prime and was beginning to sink lower and lower.

Although the second 500-year cycle saw the emergence of no great powers in the West, the life and teachings of Jesus had begun another kind of revolution and transformation. During the next cycle, while the Roman Empire was crumbling, the spread of Christianity emphasized the importance of the individual life and conscience and the universal brotherhood of man. The Roman Empire struggled for its very existence all during this period, and the ever-shrinking boundaries of its once far-reaching domain dissolved to memories when the cycle ended. Asia was the center of political energy in the world during the second 500-year cycle. The two major events of the West were the rise of the papacy in Rome and Constantine's adoption of Christianity as the state religion. It was Constantine, ruling from Constantinopole, who carved out the later Byzantine Empire.

Our historical epoch has not known a time of greater migrations than those which occurred in the fourth and fifth centuries at the end of the second 500-year cycle. A severe deterioration of world climate, climaxed by long droughts, completely desiccated the uplands of Asia. The civilizations of the northern Mediterranean shores soon felt the force of these mighty Asiatic hordes.

The year 575 B.C. had centered in a cold-dry period which lasted for about one hundred years. The cold-dry period which began about the time of Christ's birth lasted for seventy-five years. The immediate cold period surrounding A.D. 460 lasted for only about thirty years; but for a considerable period of time before and after 460, the warm phases were very short and were interrupted by unusually long successions of cold years with abnormally low rainfall.

Long cold periods are always characterized by revolutions, civil wars, and weak governments. Often, during the cold periods, there may be no government at all in large areas of the world. So it was in A.D. 460 that empires all over Earth collapsed; trade routes were abandoned; learning declined; darkness descended upon the beacon lights of civilization. In the fifth century slavery as it was known and practiced by the ancient world came to an end and was replaced by serfdom. Although in one sense only a higher form of slavery, serfdom permitted numerous freedoms which the old forms of forced servitude did not include. When the Roman Empire collapsed altogether, cultural initiative shifted from the West to the East. An old world died painfully at the end of a 1,000-year cycle.

The rebuilding in the third 500-year cycle of our epoch was slow and the emergence from the Dark Ages difficult. The first great efforts were realized in the founding of the Greek Empire at Constantinople by Justinian and the simultaneous rebirth of Persia. In the latter part of the eighth century Charlemagne welded the Franks into the mighty state which would one day be France, but the West largely lay dormant during this cycle.

While a Dark Age engulfed most of Europe, Turks, Tartars, and Huns built one large empire after another in Asia. As part of the awakening in the world as a whole, Mohammed in 622 founded another of the world's great religions. Under the impetus of this new force, the Arabs began to extend their borders far to the East and to the West. With little resistance, they marched across North Africa. They claimed Spain, and had it not been for the strength of the Franks, they could have overrun all of western Europe. Charles Martel forced the Arabs to curtail their conquering expeditions at Poitiers in 732.

While the great Charlemagne was forging his people into a new mold of organization and solidarity, Baghdad was harboring a Golden Age of Arabian civilization, which preserved the knowledge of the ancient Greeks for the world. But by the ninth century, both empires were in a state of decay. The onset of long cold-dry periods once again brought about civil wars and migrations. Asia declined in political initiative, but not in cultural innovation. Justinian and the great Byzantine Empire, together with a vital Arabian civilization which migrated from North Africa into Spain, built a new West. At the same time, China and Japan underwent an important cultural renaissance. Baghdad entered a Golden Age. Charlemagne carved out a dynamic kingdom in Europe. But the close of the 500-year cycle saw all of these civilizations in decline. In the ninth and tenth centuries great storms and bitter cold devastated Asia, driving hordes of Northmen out of Scandinavia until they had overrun Europe down to the tip of southern Italy and had conquered England.

With the ending of the medieval age, Asia once more declined, and the 500-year cycle that would belong to Europe brought with it the Renaissance. The new world which had been born out of the furious civil wars and the invasions of migrating hordes witnessed the following great events of the fourth 500-year cycle:

- The racial pattern of England was determined first by Saxon and Scandinavian invaders, then by Normans.
- The various tribal cultures of Europe accepted Christianity, built cities, accumulated wealth, and laid the foundations for the principalities which were to become Europe's modern nations.
- Christian principalities took root in northern Spain.
- The agency of the Christian Church unified Europe.
- The revolution which began in the tenth century brought into being the modern merchant class and eventually the free, and at first democratic, city-state. The merchant class instituted the principle of justice to supplement the principle of loyalty in the feudal world, and it did a great deal to demand the inception of constitutional government.
- Crusades were carried out against the Seljuk Turks.
- There was a great revival of learning and the creation of large universities with the rediscovery of ancient Greek culture. Commerce revived and great banking systems emerged. Feudalism developed with its own economic and political pattern.
- The Holy Roman Empire was founded. In Asia there emerged the extraordinary Mongolian empire of Genghis Khan, which stretched from the Chican Sea to the borders of Europe, and which was later revived in eastern Asia by his grandson, Kubla Khan. At the close of the 500-year cycle, Tamerlane would restore the great empire in western Asia. Elaborate civilizations developed in Indochina, in Central America, and in the mountains of Peru.

Although the accomplishments achieved during the fourth 500-year cycle were many, by the time it drew to a close, the peasant classes had begun to revolt against both the authority of the church and their feudal lords. In turn, the kings and princes were warring with the popes and the church for secular power and freedom of action.

Intellectual and religious freedoms grew in spite of the Inquisition. They prepared the way for modern constitutional governments, in spite of the temporary power of the principle of the divine right of kings. They made possible naturalistic, modern science, increased the scope and opportunities for self-expression by demanding communication through nationalistic tongues, instead of Latin; and at the same time brought an undreamed-of freedom of spirit into the music of the Western world. A new and free merchant class had come into being, and emerging banking systems were encouraging an industrial, as well as a cultural, revolution. The great Asiatic empires had collapsed. The magnificent cities of Indochina and the empire of the Maya in Central America succumbed to the ravages of an encroaching jungle. The kingdom of the Peruvian Incas sunk into decline. As 100-year, 500-year, and 1,000-year cycles all ended at the same time, the old world died in a cataclysm.

As part of the new awakening, the Renaissance burst forth. For the first time in our historical epoch, large numbers of people began to express themselves in art, in literatures of their native tongues, in bold explorations to faraway places. New universities sprang up in an intellectual climate much more conducive to free expression. The Reformation, with its emphasis on religious freedom, gave vital impetus to the onward march of democracy. The dawn of modern science established a social revolution based on the substitution of natural for divine laws. The invention of movable type printing only stimulated these multirevolutions.

As it became warmer on the 1,000-year cycle, the great centers of progress moved north, where they have remained to the present day. The first modern nations to acquire strength were Spain and Portugal, which upon their decline were replaced by England, France, and Holland. Austria assumed the position of power left vacant by the Holy Roman Empire. A new mercantile system emerged which promoted colonization, commerce, and industry. Absolute monarchies gave way to constitutional governments. New and independent nations were established in the Western Hemis-

Figure 14 WORLD CULTURE CURVES

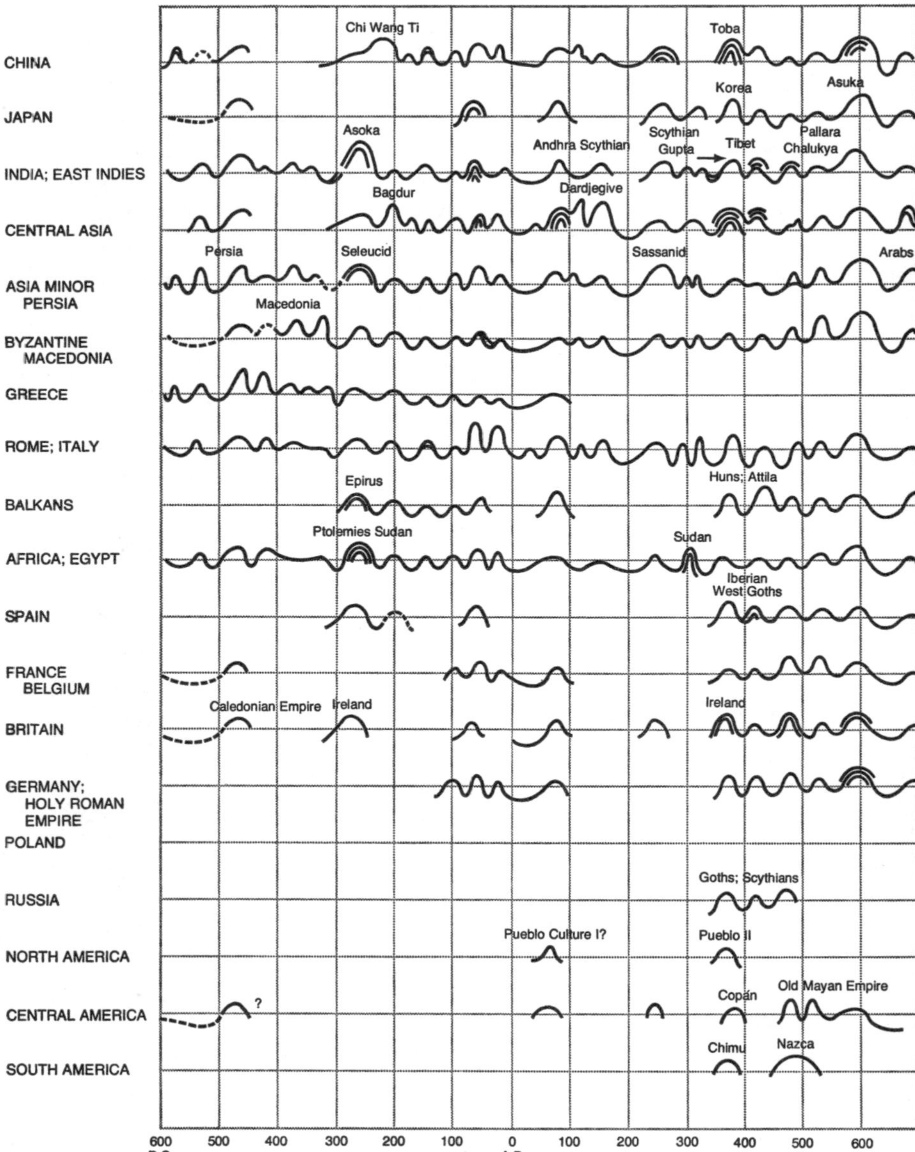

Ascents in these world culture curves indicate Golden Ages and empire-building periods. Descents are periods of decline, and "valleys" are periods of civil war. These peaks and valleys correspond to the warm and cold periods of history. Note the worldwide correspondences and universal acceleration at 1100 and again at 1500, which also correlate with an acceleration in climate fluctuation as measured in sequoia tree rings.

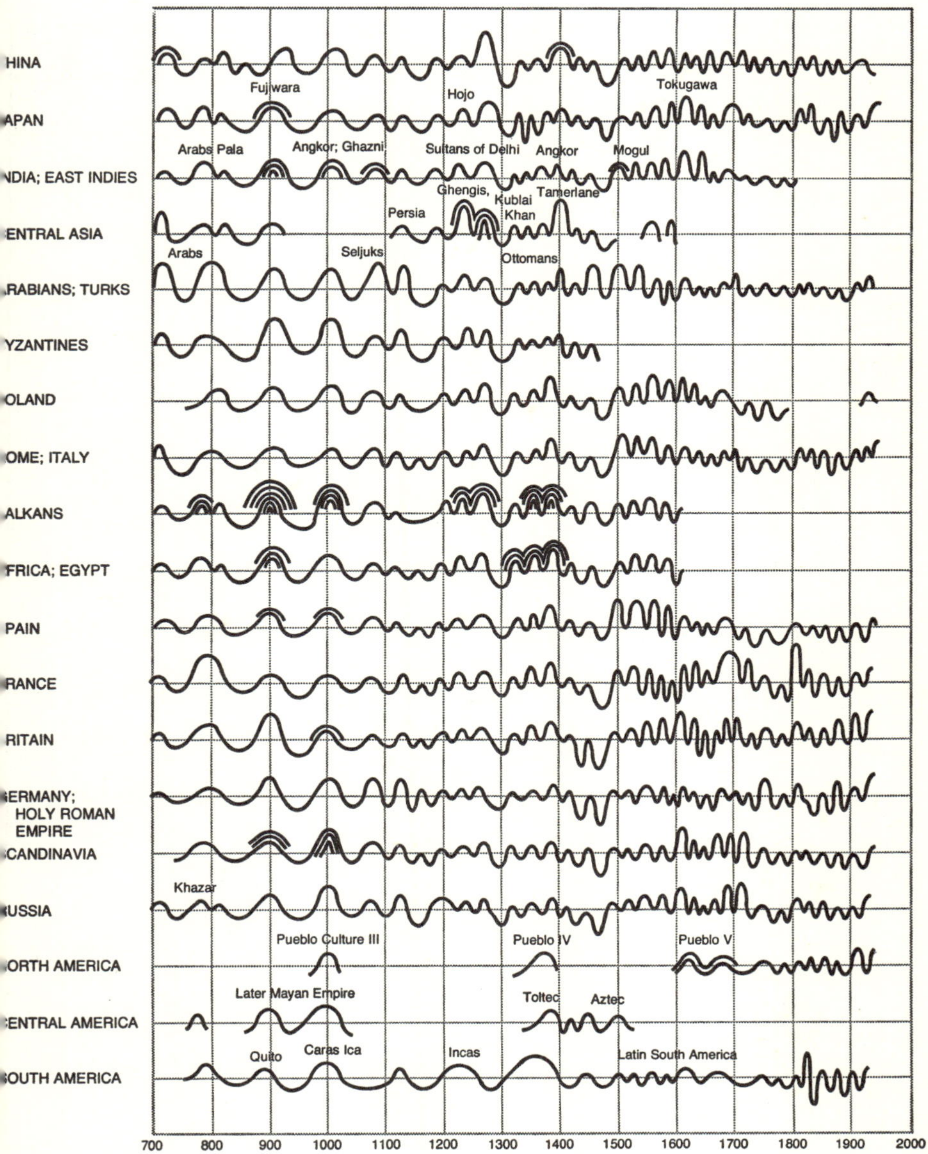

HINA
APAN
NDIA; EAST INDIES
ENTRAL ASIA
RABIANS; TURKS
YZANTINES
OLAND
OME; ITALY
ALKANS
FRICA; EGYPT
PAIN
RANCE
RITAIN
ERMANY; HOLY ROMAN EMPIRE
CANDINAVIA
USSIA
ORTH AMERICA
ENTRAL AMERICA
OUTH AMERICA
Fujiwara
Hojo
Tokugawa
Arabs Pala
Angkor; Ghazni
Sultans of Delhi
Angkor
Mogul
Ghengis,
Kublai Khan
Tamerlane
Persia
Arabs
Seljuks
Ottomans
Khazar
Pueblo Culture III
Pueblo IV
Pueblo V
Later Mayan Empire
Toltec
Aztec
Quito
Caras Ica
Incas
Latin South America
700
800
900
1000
1100
1200
1300
1400
1500
1600
1700
1800
1900
2000

phere. The United States, a child of Europe, became the powerful equal of its parents.

Wheeler's Ratios for Determining the History of Earth's Climate

As I studied Raymond Wheeler's notes, I came across a fascinating section in which he explains how he derived a "simple, workable series of ratios by means of which to represent the history of the Earth's climate." Because this data may be of interest to readers who wish a more detailed explanation of the mathematics of cyclical research, I quote Wheeler's notes verbatim:

"Since a study of cultural phenomena was responsible for this project and for the basic dates that led to the original climate curve, it will be important to explain how these dates were selected.

"The dates 350 B.C., A.D. 1250, 1820, and 1930 were originally defined as marking the climaxes of five successive empire and dictatorship periods, each intervening period shorter than the one preceding. The dates A.D. 450, 1450, the mid-eighteenth and mid-nineteenth centuries, were then interpreted to mark the centers of revolutionary and reconstruction eras. It looked as if these dates, mathematically considered, were positive and negative maxima on an accelerating cyclic curve and, therefore, were mathematically related. This proved to be the case.

"The dates 350 B.C., A.D. 1250, 1650, 1817, and 1928 are related in a very precise way. The interval 1817 to 1928 is 111 years long. The interval 1650 to 1817 is 166 years long. The first interval is 8/12 of the second one. The third interval, 1250 to 1650, is 400 years long—166 (166.5) is 5/12 of 400. The second interval, going back, then, is 5/12 of the third. The third interval, of 400 years, turns out to be 3/12 of the 4th, the interval from 350 B.C. to A.D. 1250, which is 1,600 years long.

"The series of intervals 3/12, 5/12, 8/12, may be written $3^2 5^3 8$ where the ^{2} indicates the difference between 3 and 5, and the ^{3}, the difference between 5 and 8. This looked like a series that went on, thus, $2^1 3^2 5^3 8$.

"Two twelfths was then assumed to predict a preceding interval of which the 1,600 interval was that fraction. The inferred fifth interval of 9,600 years goes back to the date now generally assigned to the climax of early Neolithic culture.

"The appearance of the series $2^1 3^2 5^3 8$ suggested a difference number of 0 in front of the 2, which in turn suggested another 2, thus, $2^0 2^1 3^2 5^3 8$. Then, of course, it was apparent that the series reversed in this fashion $8^3 5^2 3^1 2^0 2^1 3^2 5^3 8$. The next step was to perform the calculations.

"It was assumed that the interval of 9,600 years between 9,950 and 350 B.C. was 2/12 of a preceding interval, number six, which, calculated, is 57,600 years long, with its lower end at 67,550 years B.C. This date was encouraging, since it struck one assumed climax of Mousterian culture.

"Going on, the next interval, the seventh of which the sixth is 3/12, is 230,400 years, terminates at 297,950 B.C., a good date for the climax of Acheulean culture. The eighth interval of 552,960 years terminates at 850,910. No stretch of the imagination is required to assume an early Paleolithic culture, perhaps pre-Chellean, flourishing at about this time. The ninth, and last, interval turns out to be 829,440 years long, ending about 1.7 million years ago, a convenient date for early Eolithic culture.

"Now if it be assumed that these dates are maxima on one side of an axis, and that there are minima (negative maxima) on the opposite side of the axis midway between them, one obtains the following:

"In the center of the sixth interval, 38,750 B.C., one finds the climax of the fourth of Wisconsin-Würm Ice Age.

"The center of the seventh interval is 182,750 B.C., which strikes the average estimate for the third, the Illinoian, or Riss, Ice Age.

"The center of the eighth interval is 574,330, an average date for the second ice age, the Kansan-Mindel.

"Finally, the center of the ninth interval lies near certain of the more recent estimates for the first or Nebraskan-Günz Ice Age, that is, 1,265,630 years ago. (These figures are carried clear out merely to follow through the results of the calculations.)

"Thus the cycles seemed to be assuming definite physical properties, for represented on one side of the axis were prehistoric human cultures that developed during warmer interglacial periods, and on the other side were the four successive glacial periods of the Pleistocene (assuming the Wisconsin-Würm series to be represented by its climax)."

Wheeler goes on to relate how a corresponding series of alternating warmer and colder periods were found for geologic time, even though the convenient series of twelfths appeared to have given out.

A number of standard texts of historical geology enabled him to conduct empirical study of the Paleozoic, Mesozoic, and Cenozoic eras and to find justification for his assumption that certain epochs had centered in warm times while others had been cold, mountain-building times, "falling conveniently midway each time on the other side of the axis." Additional research revealed that here, too, existed another series of twelfths, leading in alternate colder and warmer periods right back to the date 8.321 billion, "for a possible warm maximum near or at the origin(?) of the Earth."

"Two series of fractions, both in twelfths, both of the same length, and both having the same series of 'difference' numbers, or lag and lead numbers—3,2,1,0,1,2,3—seemed to 'predict' a continuous series of *mathematically related accelerating* cycles that started at the beginning(?)of the Earth and came down to the present time," Wheeler writes. In going over his computations, Wheeler was intrigued that both of his series of fractions indicated that there had been an unusual stepdown by the same fractions, twelfths (upon which both series were based), which had occurred during a "gap or joint" between 1.8 and 10 million years ago.

"At the time of this stepdown, a peculiar place mathematically in the total series, something unusual happened in the history of Earth—the rise of the midcontinental Himalayas, Alps, Rockies,

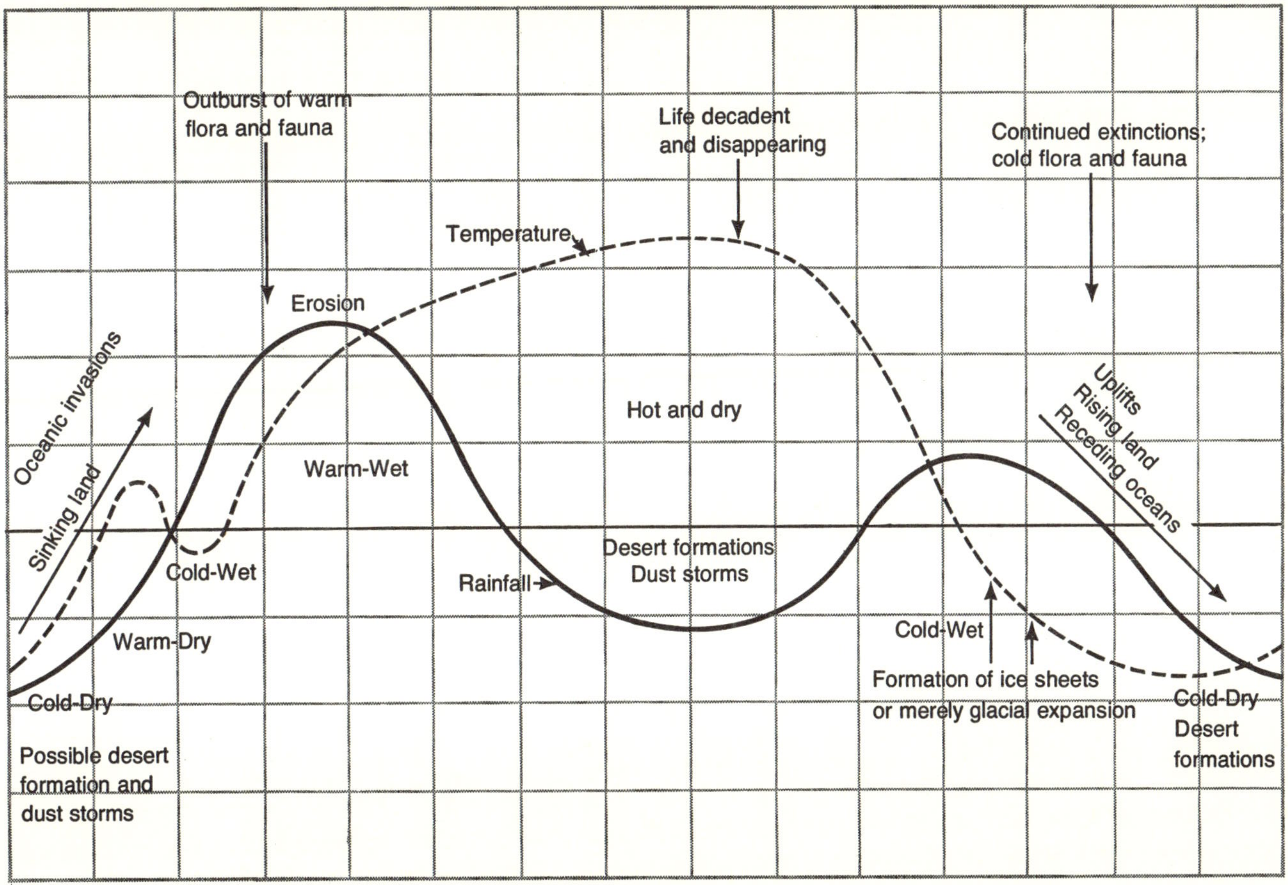
Outburst of warm flora and fauna
Life decadent and disappearing
Continued extinctions; cold flora and fauna
Temperature
Erosion
Oceanic invasions
Sinking land
Hot and dry
Warm-Wet
Uplifts
Rising land
Receding oceans
Cold-Wet
Desert formations
Dust storms
Rainfall
Warm-Dry
Cold-Wet
Cold-Dry
Formation of ice sheets or merely glacial expansion
Cold-Dry
Desert formations
Possible desert formation and dust storms

perhaps the greatest of all mountain-building periods," Wheeler noted.

Wheeler was left to grapple with more recent dates—and the question of what had been happening since 1928 and what would happen in Earth's future. The 8,5,3,2,2,3,5,8 series had ended at 1928. Was there another interval, another gap or joint occurring now like the one between 1.8 and 10 million years ago?

"It takes no imagination to see that another stepdown like the one during the Pliocene could not occur geologically and mean anything because it would be too short. It is incredible that the cycles should cease altogether. The only reasonable conclusion would seem to be that the cycles will *step up,* that is, *reverse.*"

Wheeler's extensive research of the time cycles suggests such a reversal of the cycles has occurred three times in the past—and profound evolutionary events have occurred during each cyclic fluctuation. During the first reversal, "the vertebrate pattern emerged and stabilized sufficiently to make possible the profound modifications and differentiations of structure compelled by the more strenuous fluctuations of the Mesozoic Era."

Had the vertebrate pattern not stabilized in relation to environmental forces, Wheeler speculates, it might well have been extinguished altogether during this era.

During the second climatic optimum, Wheeler tells us, another profound step in evolution occurred—the emergence of the mammal and primate patterns and the beginning of prehuman forms. "Thus an enormous step now occurred in evolution, comparable to the emergence of the primitive vertebrate during the Silurian interval."

Then, writes Wheeler, "The primate pattern had developed sufficiently by the end of the last ice age to permit a *third* profound occurrence during that series of rapid and mild fluctuations. This profound occurrence was the emergence of modern man and his psychological and social achievements." Purely as a heuristic principle, Wheeler points out, the deceleration-acceleration theory of the history of Earth's climate is worth further investigation. "There are possible implications that have an important bearing upon human problems today," he suggests. "This would mean that a new and probably different series of similar general form will soon begin, and thus Earth is about to begin a new phase of its history."

14

Business Trends in the United States, 1790-1900

There has never been a satisfactory economic theory of the business cycle for the simple reason that no such theory can exist. Business cycles are not uniquely, nor even basically, economic. It is a mistake to think of business cycles exclusively in terms of current phenomena such as modern banks, capitalism, stocks, and the like. These are all passing and incidental, while periods of prosperity and depression are eternal.

Business Week (July 20, 1974) recalled the work of Nikolai D. Kondratieff, a 1920s Russian economist, who discovered a fifty-year price cycle. "According to Kondratieff's correlations, the 1970s may be the Roaring Twenties all over again. Specifically, 1974 seems analogous to 1920. Inflation is almost ended. There may be a recession in 1975, but the Great Bull Market is ahead . . . the Great Crash won't come until the early 1980s, after deflation has sunk its teeth into the entire economy."

Kondratieff was sentenced to Siberia because his economic fatalism was not compatible with Marxist dogma. Kondratieff espoused no ideology, however, and he offered no philosophic explanations to account for the price cycles; he simply observed patterns.

Whether there is a consistent relationship between business and weather trends is not an isolated problem. It is but a segment of the larger question of the relationship between human behavior in general and the weather trends. As Dr. Raymond Wheeler stated:

"Not a single, basic aspect of human behavior and productivity escapes from its relationship with the weather trends. Trends in government; scientific, economic, philosophical thinking; invention; painting; music; design; literature; education; religion; all shift predictably with the phases of the 100-year climatic cycle. Were business trends exempt, it would be fantastic; but they are evidently *not* exempt.

"Stated deductively, it is obvious that whatever explains the relationship of the human behavior pattern as a whole to weather trends must also be the basic explanation for whatever relationship is found between business and the weather.

"Stated inductively, whatever relationship is found to exist between specific kinds of human behavior, studied one by one, and the weather trends, would most likely be found to hold also in connection with business trends and the weather. If the weather has any effects at all on human behavior and human productivity, it would be logical to suppose that, in general, the effect would be the same on all types of places. The facts bear out this assumption.

"The universal correlation between human behavior and the weather trends grows out of the effect that temperatures and storminess combined have upon the metabolism of the body or, in simpler language, upon human vitality, or the level of energy available for work. Good health and a high energy level are synonymous, as are poor health and a low energy level.

"In general, a combination of cool temperatures and cyclonic storminess (which are found together) is the best for human health, vitality, aggressiveness, and intelligence. Corresponding conditions produce the best crops, in the sense of the greatest yield per acre, whether potatoes, corn, wheat, rice, or other products, assuming, of course, the proper soil and agricultural care."

"Why should there not be optimum weather conditions for the conversion of man's energy, obtained from food and air, into that kind of work we call business?" Dr. Wheeler asked. "Evidently there are such optimum conditions. These conditions are to be found on the two transitions of the climatic cycle, the one from cold to warm and the other from warm to cold."

According to the evidence accumulated by Dr. Wheeler, for the best human health, the highest intelligence, and the most aggressiveness, there should prevail the following:

1. Mean annual temperatures from 47° to 52° F., inclusive, for the very best. The range 45° to 55° is good; 40° to 60° fair; and 35° to 65° possible.

2. There should occur the cyclonic type of storm, producing rainfall twelve months in the year. There should be an annual temperature range of not less than 60° and not more than 160°. (New York has a range of 110°; Winnipeg, 150°; and San Francisco, 65°.)

3. There should be an average daily range of 10 to 30°. Wider ranges than these are possible, but only within limits.

In the charts which follow, a comparison has been made between the weather trends of the United States since 1794 and the business cycle as measured in terms of business volume. Based on the past performance of temperature and rainfall in the United States from the year 1794, Dr. Raymond Wheeler drew a theoretical curve for each of the two climatic variables and compared it with actual records of temperature and rainfall. The comparison may be seen in detail in the charts accompanying this chapter. Each curve is composed of six cycles. The last times of occurrence of the maxima and minima of these cycles determined their position throughout the curves.

How could Dr. Wheeler know how warm, cold, or dry it was

in the United States between 1794, when Chart I begins, and the founding of the United States Weather Bureau in 1870?

First of all, there do exist a few long temperature and rainfall curves which extend back into early American history, kept by early settlers and officials of colonial and local governments.

Then, too, one must remember that at one time Dr. Wheeler had as many as two hundred researchers reading widely in order to supplement scanty records. Accounts of droughts, floods, crop failures, early and late frosts, heat and cold waves, blizzards, grasshopper plagues, and the like, massed together, aid enormously in locating the positions of the various phases of the climate cycle. Such reports come from official records; chronologies; private diaries, documents, and manuscripts of various kinds; and early periodicals and newspapers.

Tree-ring curves are available from many sections of the country, from the New England states south to Georgia, and west across the Mississippi Valley through the Rockies to the Pacific Coast. Western data extend from Alaska south to the giant sequoias of California and the pines of the Southwest. It can be demonstrated that the story of weather trends, as told by the trees in the widths of their annual rings, tallies very well with rainfall and temperature curves, as far back as the latter go.

Continuous data on sunspots go back to the middle of the eighteenth century. Students of climatology and meteorology accept the fact that temperatures rise on the approach to a sunspot minimum and fall on the approach to a maximum.

Rainfall and temperature data in abundant amounts for the last eighty years demonstrate a very close relationship between weather trends in the United States and weather trends in the world as a whole. No trend is 100 percent uniform, either in small or in large areas; but the variability of the trends in different parts of the United States seems to be hardly more or hardly less than the variability of trends between the United States and other continental areas of the world, including Europe. The variability, as far as changing trends is concerned, seems to be approximately the same whether the area is large or small.

There is in round numbers an overall correspondence of 74 percent, on a year-by-year basis, between wet and boom years on the one hand and dry and depression years on the other.

A boom era is defined as a period on a climatic transition during which boom years outnumbered depression years. Eight of these boom eras are distinguishable in the history of the United States since 1794. With no exceptions, these were the wetter periods in United States history. A depression era is defined as an era in which, for several years, time spent in depressions exceeds time spent in periods of prosperity. By several years is meant any period of time from five to twenty years in length. These periods are identified by their positions at the crests and troughs of the temperature curve. There have been six conspicuous periods of this sort in United States history since 1794, four major and two minor. The major occur on the main 100-year cycle, the minor on secondary cycles. With one exception these eras have occurred during times that averaged dry. The one exception was a minor period in which it was severely dry half the time in the late 1870s.

The highest correspondence occurred between boom and cold-wet years; i.e., 85 percent.

The next highest correspondence was between boom and warm-wet years; i.e., 78 percent.

Correspondence between warm-dry and depression years was 75 percent.

Correspondence between cold-dry and depression years was 56 percent.

The 56 percent is not the exception to the rule that it appears to be, because many studies and miscellaneous information pertaining both to man and to animals show that cool temperatures are themselves invigorating, regardless of rainfall. A poorer correspondence between cold-dry years and depressions is therefore to be expected. On the other hand, when cold-dry years during which business sharply dropped off are not defined as exceptions, the correspondence rises to 66 percent.

When, for the "rule" to hold, the criterion is taken that the majority of the years occupied by a boom should be wet and the majority of the years occupied by a depression should be dry, the correspondence between all wet periods and booms, and all individual dry periods and depressions, is 85 percent. In other words, during a short wet period, a boom can be expected at least about 8.5 times out of 10, and likewise with depressions and dry periods.

1791-1800

The cold period of the 1780s was a prominent part of the cold phase of the 100-year cycle and was definitely worldwide in scope, as was the return to the warm side in the early 1790s. The rainy period of the early 1780s, while it was cold, was likewise worldwide—as was the dry period that began in 1785. One of the obvious causes which precipitated the French Revolution of 1789 was the fact that crops had been poor, and it had been very cold. The year 1789 was cold and dry in France; and because of crop failures and suffering during the winter, the bitterness of the people against an oppressive and arrogant nobility had been greatly increased. The revival of rainfall in 1790 was worldwide, and the dry years in the late 1790s occurred over large and widely separated areas of Earth.

A great boom and industrial revolution occurred in the 1790s (and in the early part of the nineteenth century, to about 1818) during nation-building time, on the transition from the cold to the warm side of the 100-year cycle, and into the warm-wet phase of the cycle. The warm-wet phase of the 100-year cycle began early in 1794. At this time there was a great renaissance in literature, art, and science—one of the most brilliant epochs in history. All human behavior, including business, demonstrated that man was living at a very high energy level.

"Seaborne commerce prosperity," as the boom of 1794-1796 is conventionally called, would undoubtedly have occurred in the absence of any particular economic incentive. Weather trends, especially rainfall, promote incentive of increasing human vitality and aspirations.

Figure 16 PREVAILING WEATHER TRENDS IN THE UNITED STATES 1791—1800

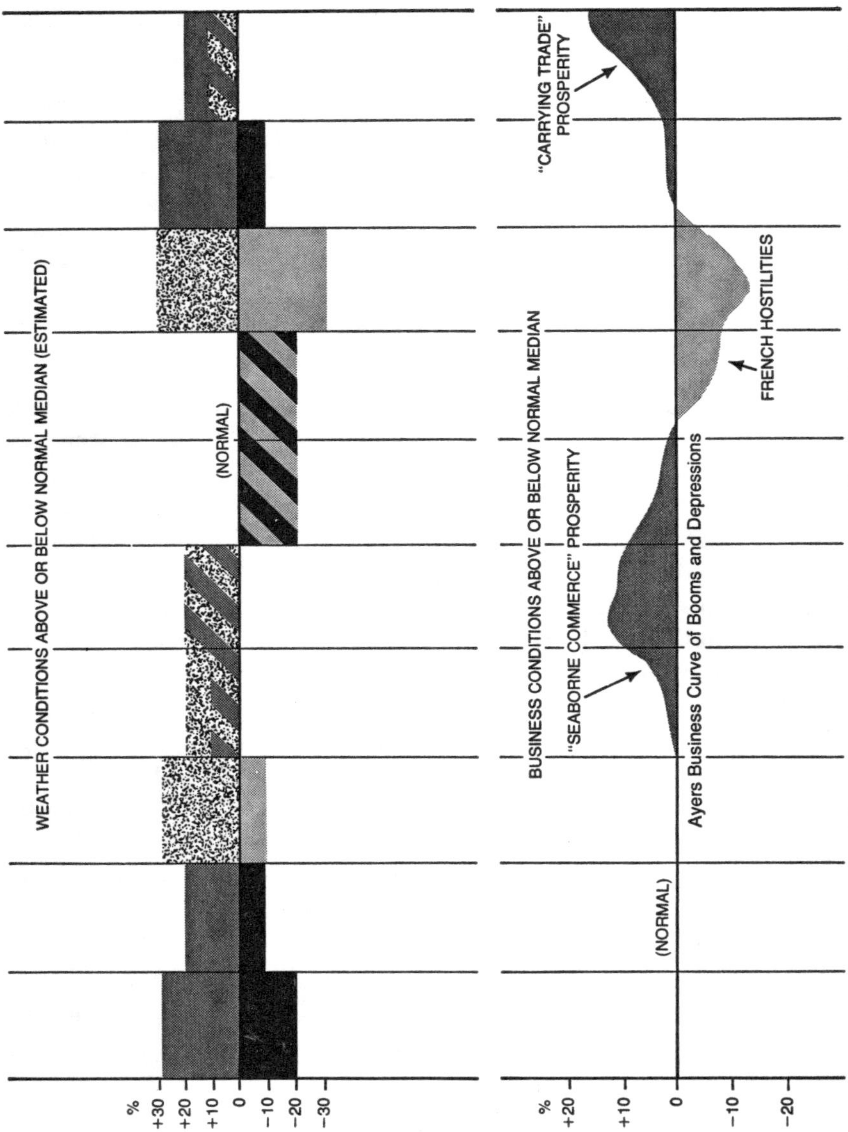

In this chart and following:

represents Warm

represents Dry

represents Wet

represents Cold

Where stripes appear, this means that the period illustrated was a mixture of two conditions (for example, 1794 and 1795 were Warm and Wet). The Ayers curve of business conditions, below, has been filled in to emphasize the correspondences.

1801-1810

The years 1810 and 1811 were the last of the warm-wet years. It is interesting that during the dry spell, 1801-1803, there was a lull not only in the Napoleonic Wars, but in warfare all over the world. Just as the dry wave was worldwide, so was the wet climax at 1810-1811, and the cooler years, 1808-1809.

The boom that began in 1790 lasted over into dry 1801, but after that the dry years took hold. The boom stopped as dry years set in. While the depression was attributed to French hostilities, the chances are overwhelmingly high that, whatever the economic reason, the depression would have occurred anyway.

Business, commerce, and human activity, including warfare, picked up as it turned wet again. Business volume fell during dry 1806, but the boom had enough vigor to recover and reach its climax during wet 1807.

In 1808 the dry year was associated with a depression. When the dry period is short, so is the depression. Weather trends are more basic than economic factors, because they predispose the economic causes and factors to emerge, or to change, through the conditioning of human vitality and moods. The weather trends time the appearance of economic and other factors and, in many instances, explain their strengths and weaknesses.

1811-1820

Although wars do not ordinarily begin during dry years, the War of 1812 developed as a result of the determination of the War party which had been active during wet 1809, 1810, and 1811.

The cold wave from 1814 to 1818 was worldwide, and historically, a famous one. It was a civil war period the world over. Rebellions against the Napoleonic regime broke out, and the revolutions which freed Latin America erupted.

If the War of 1812 had not occurred, there undoubtedly would have been a depression at that time. Dry 1816 and 1817 did not produce much of a depression, but there was a sharp decline in business volume.

Figure 17 1801—1810

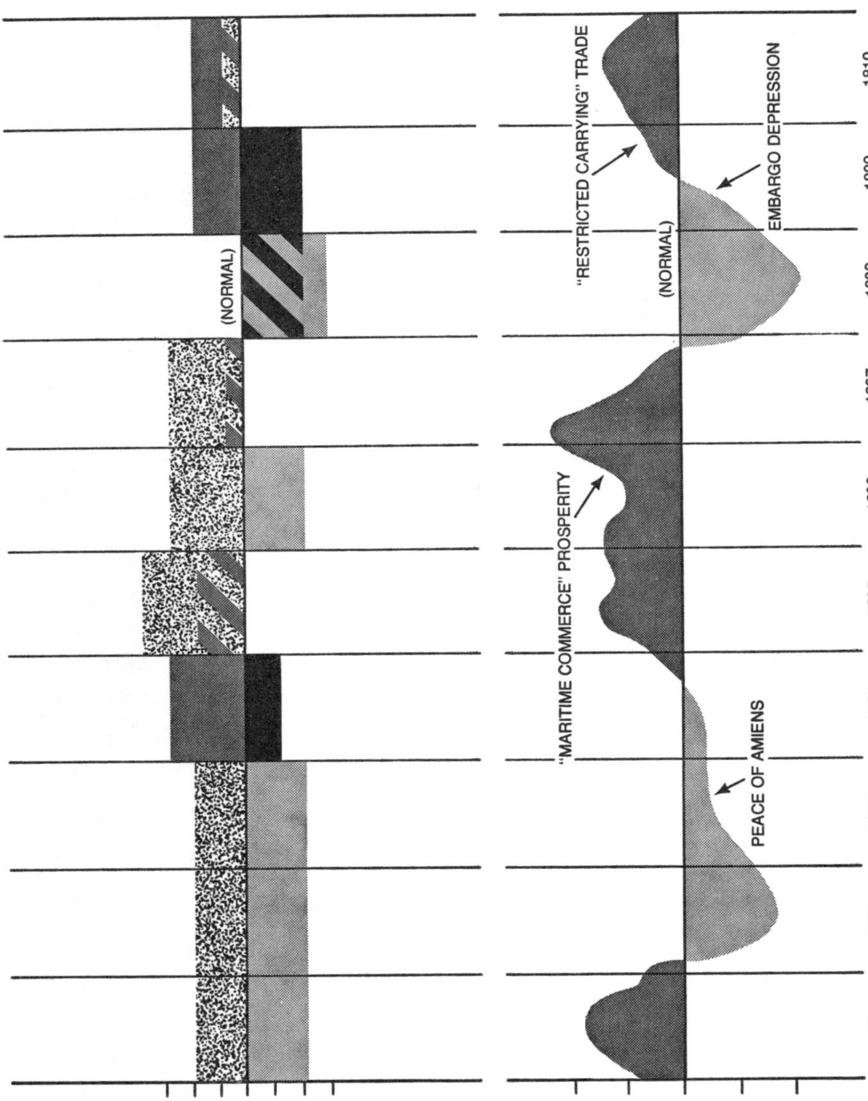

%
+30
+20
+10
0
−10
−20
−30
(NORMAL)
WAR OF 1812
%
+20
+10
0
−10
−20
(NORMAL)
WAR OF 1812
POSTWAR RECOVERY
PRIMARY POSTWAR DEPRESSION
1811
1812
1813
1814
1815
1816
1817
1818
1819
1820

The so-called postwar recovery in 1814 was a function more of wet years than the consequence of peace.

The single wet year, 1818, was accompanied by a single boom year. Then the first of the depression eras occurred during the warm-dry phase of the 100-year Cycle 26. Nine out of twelve years were spent on the depression side, and there were only three short periods of recovery during this period. The depression of 1820 is comparable to the depression of the 1930s. This was a major depression associated with a major phase of the climatic cycle. Warm-dry 1819 and 1820 were rather severe depression years. There has been a tendency throughout history for depressions to be much worse during warm-dry than cold-dry times—about 50 percent worse since 1794. Perhaps this is because warm-dry times are more debilitating than cold-dry times. The warm-dry phase of the 100-year cycle occurred in the 1820s and was conspicuous the world over.

1821-1830

Wet 1821 was associated with a business recovery which lasted for a short time into 1822. As rainfall improved in 1824 and 1825, so did business. Wet 1830, the beginning of a succession of wet years, initiated a four-year boom. When it turned cold about 1830 and rainfall revived, prosperity returned to the civilized world. The British wheat price rose a hundred points. Business shifted to the boom side.

Look ahead to Figure 20 and notice that it suddenly turned cold after 1830. The warm-dry phase generally terminates the warm part of the 100-year cycle. The warm-dry phase occurred again in the 1930s, 110 years later. The warm-dry phases of the 100-year cycle have been associated with economic decline down through history. The health and vitality of civilization in general, including the birthrate, have also declined during these periods. The great crash at the end of 1929 began just as a great drought began and continued as long as the drought lasted.

Figure 1821—1830

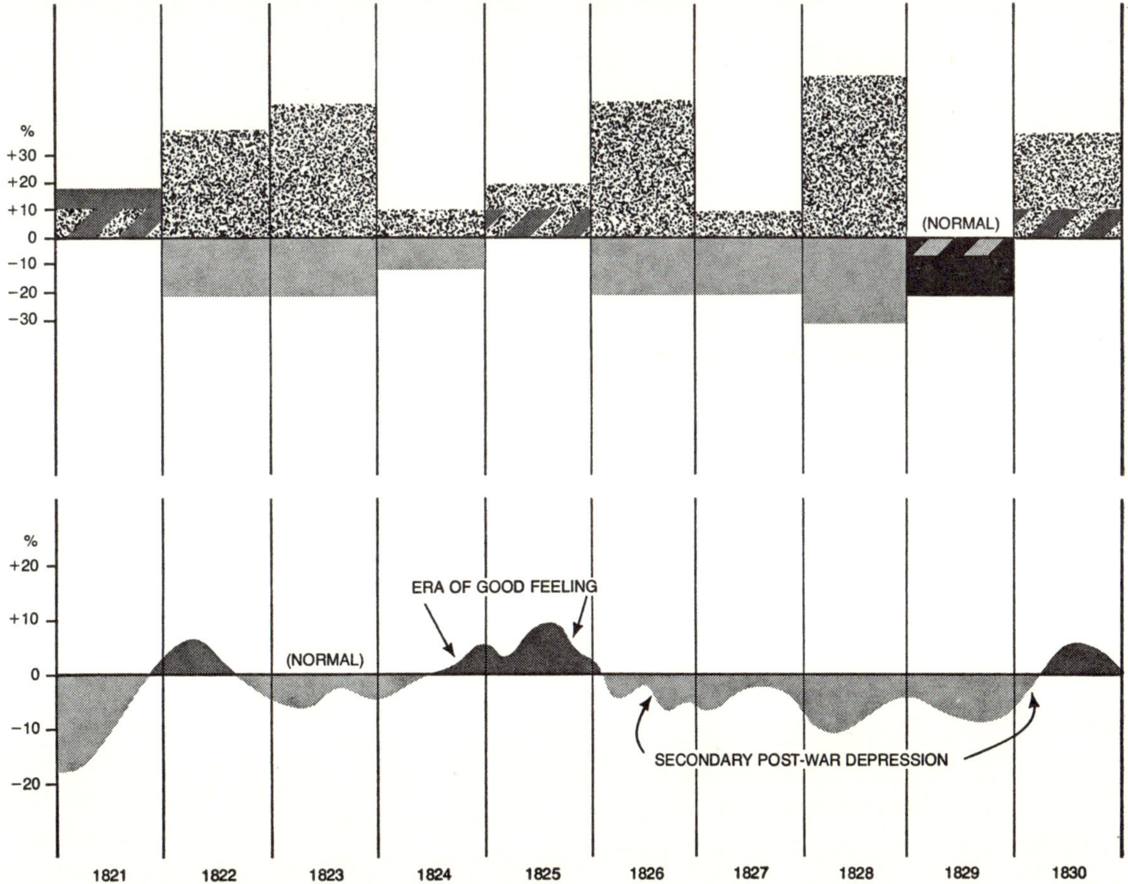

1831-1840

Cold-wet years 1831 and 1833 intervened between the warm-dry and the cold-dry phases of the long cycle. A warmer 1832 broke into the cold 1830s, which reached a 100-year low. The second boom era occurred on the reverse transition; that is, from the warm to the cold side of the 100-year cycle in the early 1830s. This boom period was the cold-wet phase of the 100-year cycle. The boom period from 1830 through 1836 (cold-wet) had six good years and one bad year.

In 1834 it turned dry suddenly, and a depression developed. However, this time the economic recovery led the rainfall recovery. On the other hand, the main part of the boom coincided with wet 1836. Then, once again, the weather became colder, and it stopped raining in 1837. The price of British wheat hit two longtime lows—one just before 1835, while the United States was in a depression, then again at 1850 and just before, while we were in another depression. This second depression era in the late 1830s and early 1840s was associated with predominantly cold-dry times. The depression period had practically eight bad years to one good one (1836 to 1846).

This was a major depression period, associated with the warm-dry phase of a short cycle; i.e., in the second half of the 1850s. The depression was not as severe as might have been expected, probably because of the recent opening of the West.

This time, economic or other factors were strong enough to cut across the weather factors; and for a time the weather and business trends ran counter to one another.

This was the first part of the cold-dry phase of the 100-year cycle. Additional parts occurred in the late 1850s, the 1860s, the early 1870s, the 1880s, and the early 1890s. It is common for the cold-dry phase to stretch out over a long period of time and to be interrupted by warm years, followed by cold-wet years (see charts including the 1850s and the 1870s). We are now beginning a long cold-dry period.

Figure 20 1831—1840

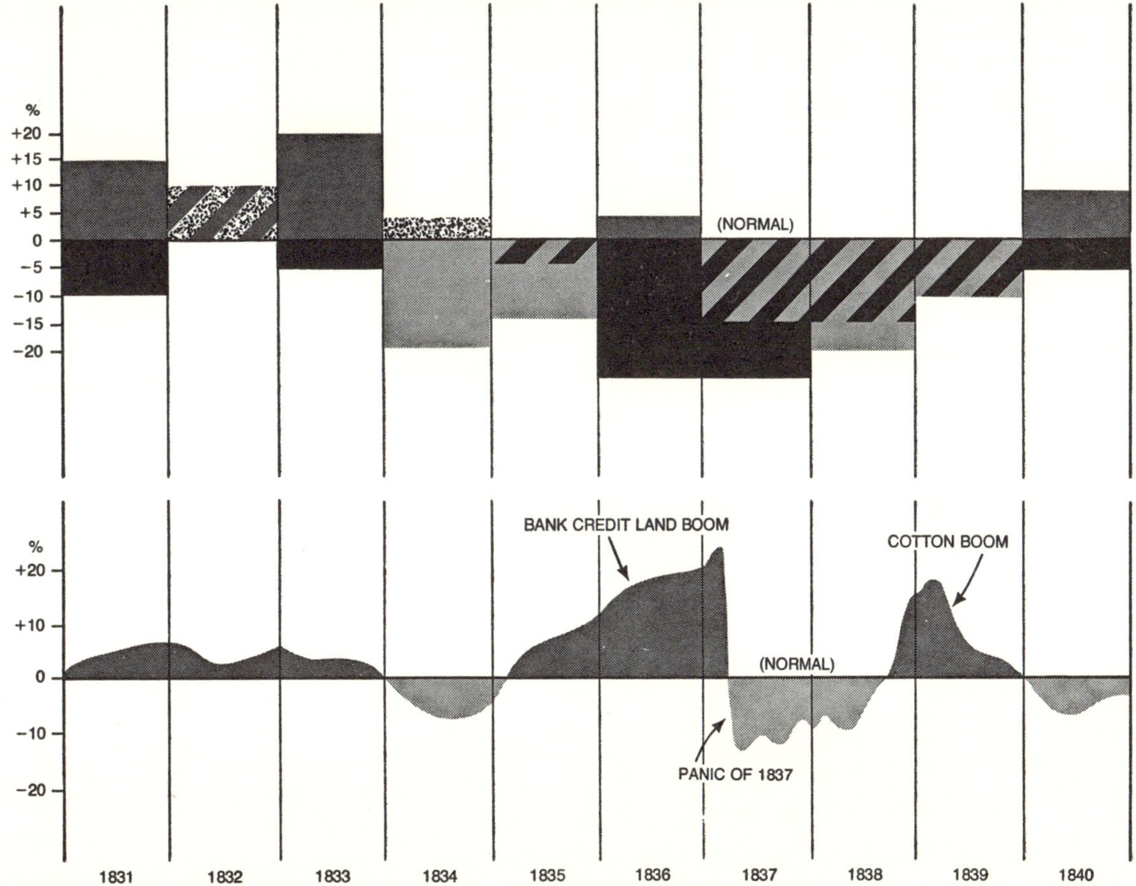

1841-1850

The cold period continued through the 1840s, and much of it was dry. The cold periods of history are eras of democratic reform, emancipation, and rebellion; and the 1830s and 1840s were known for their reforms, as were the 1860s, 1880s, and 1890s, which were also cold. Wheat prices also dipped in the early 1840s while we were having a severe cold-dry spell and a relatively severe depression.

The third boom era occurred in the late 1840s and early 1850s on a secondary cycle and on the transition from cold to warm as rainfall shifted to the wet side. Wet 1843 was associated with a rapid rise from the bottom of the depression. Rises in temperature tend to produce international wars, and on the particular rise in 1846, the Mexican War developed. The peak of the boom in 1847 coincided with a peak in rainfall, and it is almost certain that even without the Mexican War, there would have been a period of prosperity at this time. Cold 1848 was worldwide and a year of great revolutions.

1851-1860

A secondary cycle appeared in the 1850s, causing a ten-year warm interruption of the cold phase of the 100-year cycle. Then came a revival of business on a rise in temperature and rainfall. We had a similar boom in the early 1960s. As has been so often stated, rising temperatures signal rising imperialism. The Crimean War and the United States demand that Japan open her doors to trade occurred at this time.

The Gold Rush to California was a typical symptom of the aggressive drive exhibited by societies everywhere on the climatic transition from cold-dry to warm-wet periods. The drive characteristic of societies at this time is evidenced in the long boom from 1850 to 1857, which was strong enough to counteract the force of the weather moves. It is not surprising that in a young, growing nation, a boost such as was provided by the discovery of gold in California should produce a boom so vigorous that dry years had less than their usual effect.

Figure 21 1841—1850

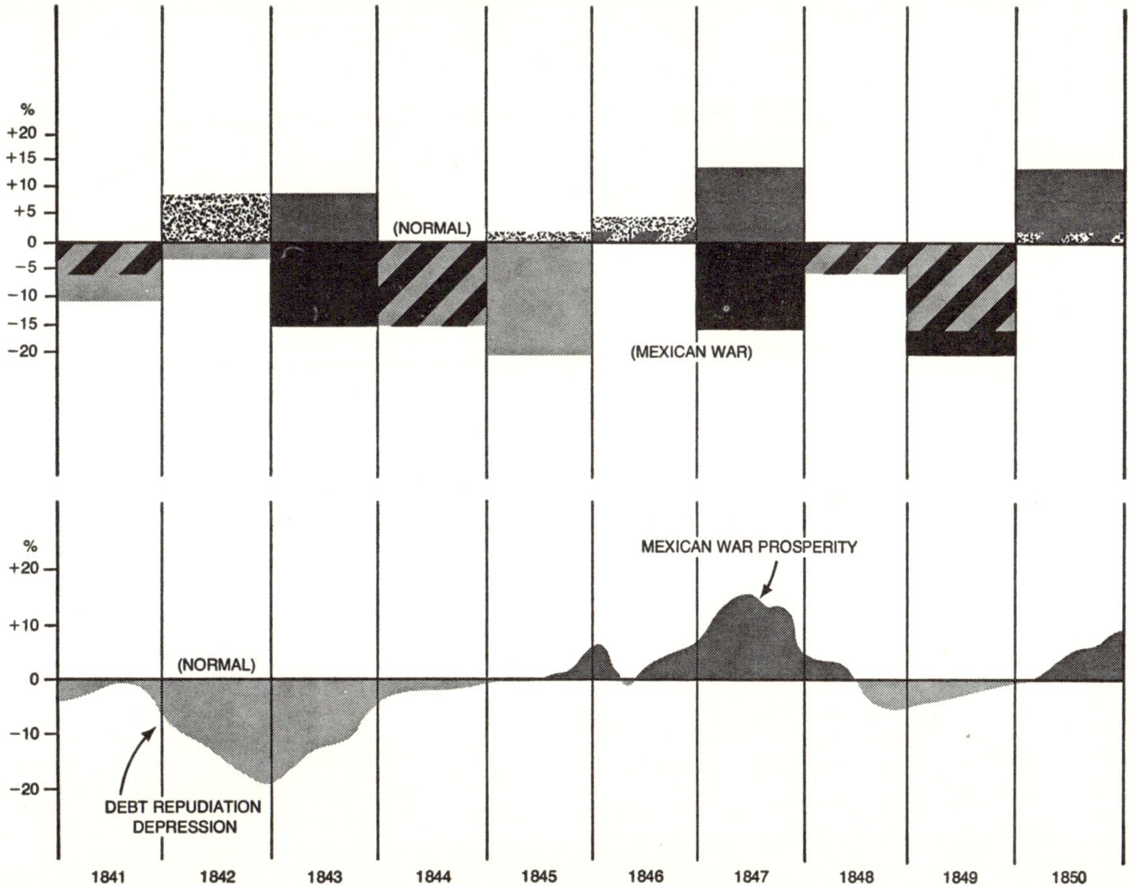

Figure 22 1851—1860

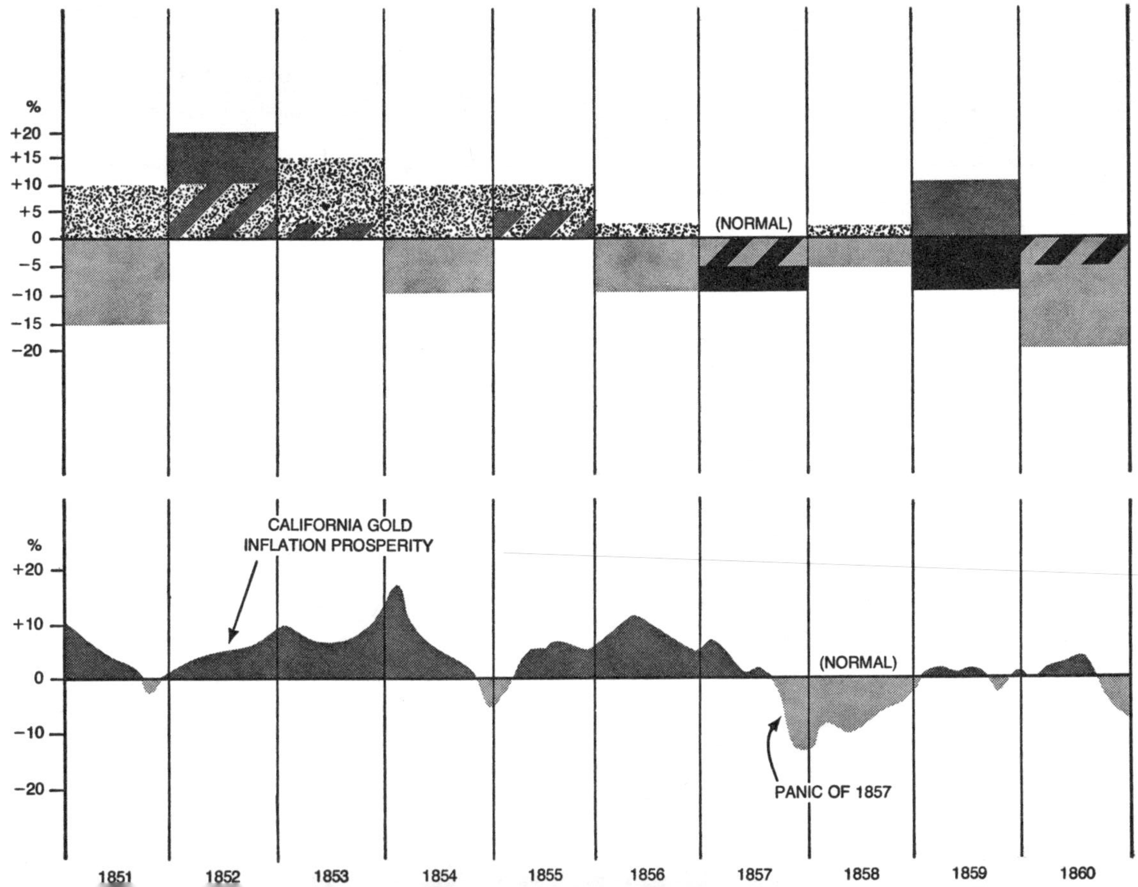

Yet notice the interesting fact that each dry year caused a drop in business volume and precipitated incipient depressions—1851, 1854, and finally a panic in 1857. This surely demonstrates that dry years had a powerful influence as effective as if the depressions had fully matured.

This fourth boom era was a period of prosperity on the reverse transition of the short cycle as temperatures shifted back to the cold side and rainfall picked up again after a lull in the late 1850s.

The end of the boom started with dry 1856. Wet 1859 brought a recovery from the panic, but dry 1860 was associated with a plunge again to the depression side.

1861-1870

Although Europe remained somewhat warmer than the rest of the world in the 1860s, the general decline in temperature brought with it civil wars throughout Earth. Cold 1859 and 1860 paved the way for the precipitation of these wars by intensifying the spirit of rebellion. The revival of rainfall in the second half of the decade was worldwide, as was the drop in temperature.

The drop in business activity in 1861 began the previous year and was associated with the onset of dryness. The outbreak of war prevented a recovery in 1861, when it was wet. The rise in business volume through 1862 was presumably a result of the war.

Notice that the sudden drop in rainfall in 1867 was associated with a tendency toward a depression, even though rainfall still averaged on the wet side. Dry 1870 produced a mild depression, which probably would have continued but for the fact that in 1871 it turned wet again.

The rise in temperature in 1870 produced another trend toward imperialism. The Franco-Prussian War broke out, together with a strenuous campaign against the American Indian tribes in the United States.

Figure 23 1861—1870

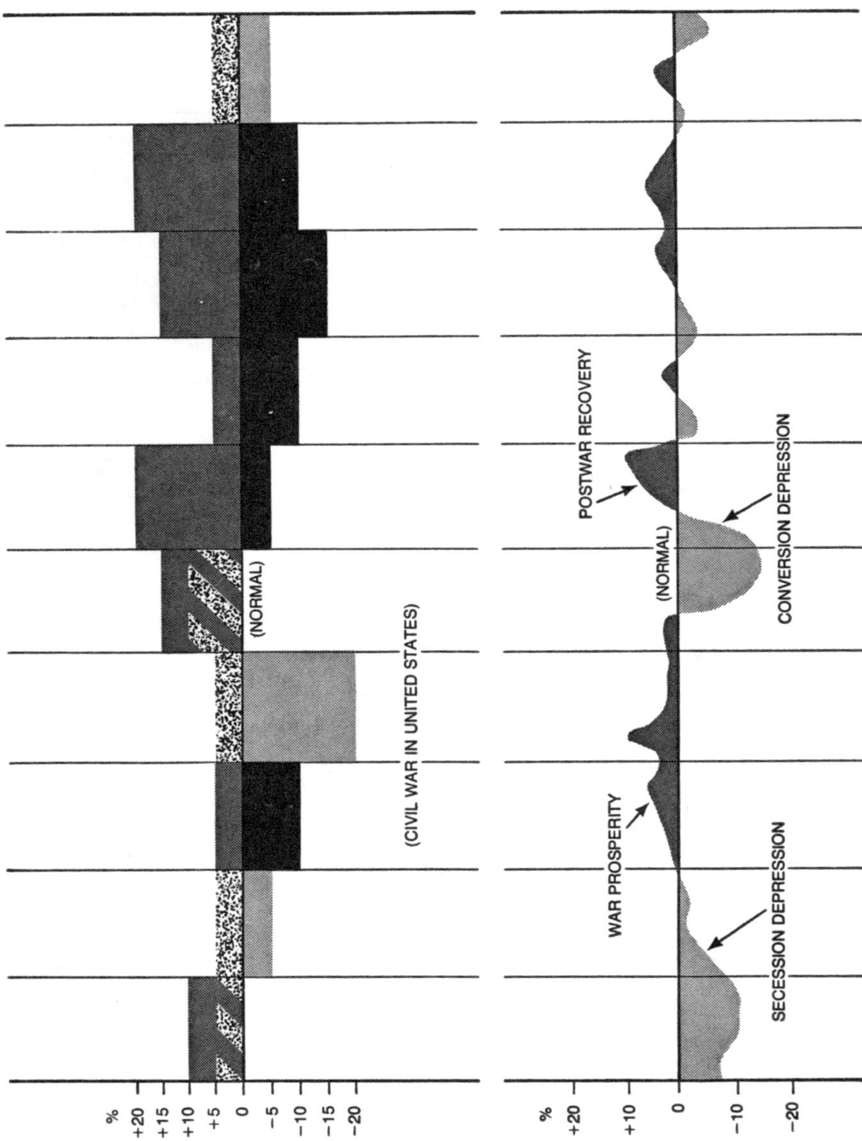

1871-1880

The cold-wet years in the early 1870s were worldwide, as was the rise in temperature and the revival of rainfall in the second half of the decade.

While there were signs of instability during dry 1872, the fact that it was cold possibly helped to maintain the boom. Again, the climax of the boom was reached during a wet year, 1873. For ten years, prosperity had prevailed, and as a consequence serious overexpansion and inflation had occurred. Had 1874 been wet, there might have been an immediate rebound, but it was dry.

The fourth slump was the long depression of the 1870s. In terms of the weather-cycle pattern there should have been a warm-dry phase at this time, in a second, short cycle. There were three very dry years during the depression, but also three wet years. This was a minor era.

The depression of the 1870s was intensified and prolonged by dryness and possible excessive heat as well. The boom of 1871 to 1873 had been a very vigorous one; and because it had gone to extremes, it called for a dramatic reaction on purely economic grounds.

Mildly wet 1875 was not strong enough to produce a recovery. Rainfall and the depression hit bottom in 1876.

Then, as rainfall revived, the depression lessened. The fifth boom era was on another secondary transition, due to another short cycle, as it turned warm again in the 1870s. Rainfall picked up again. The year 1878 was famous for its excessive heat and heavy rainfall all over the world. It was conspicuous on all five continents, even in the tropics. The rainfall of 1878 gave the recovery sufficient momentum to continue through dry 1879. The recovery was stabilized by wet 1880.

1881-1890

The weather trends turned back to cold-dry after a brief warm wave. The sixth boom era was on the reverse side of the second short

Figure 24 1871—1880

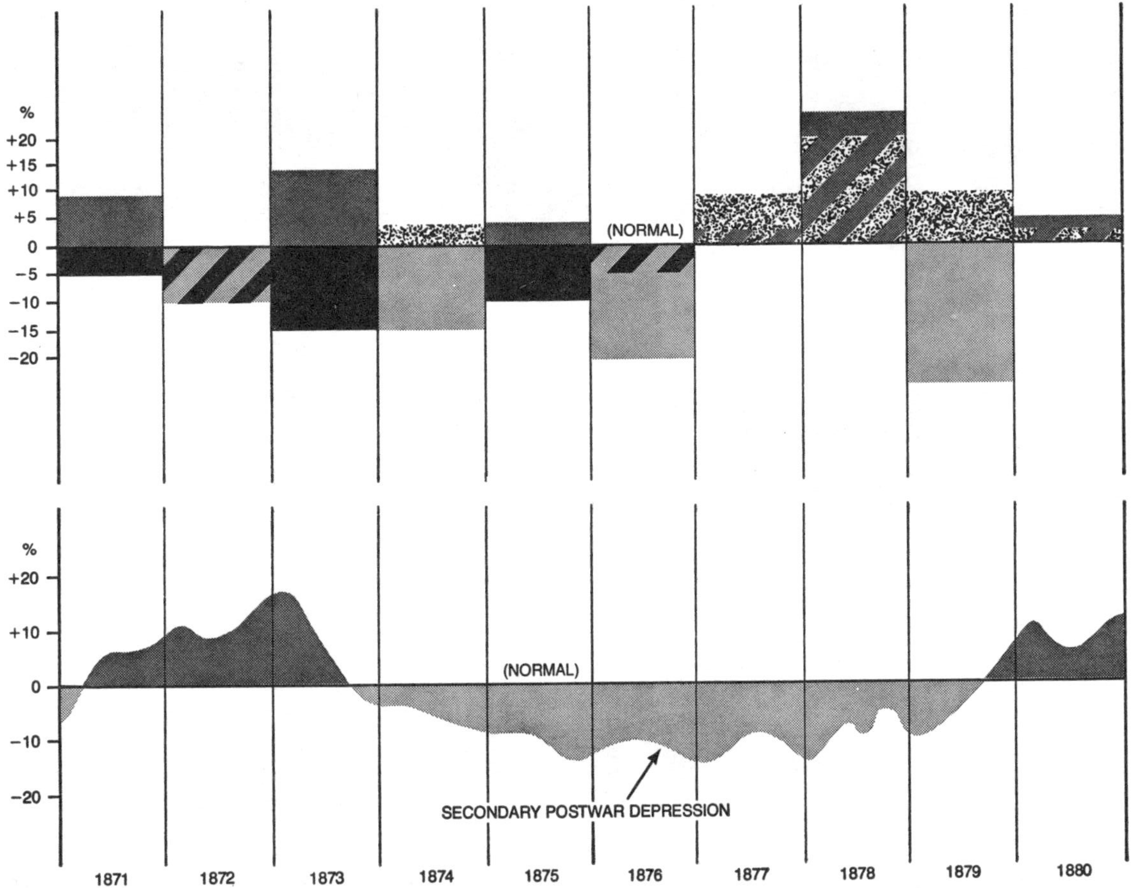

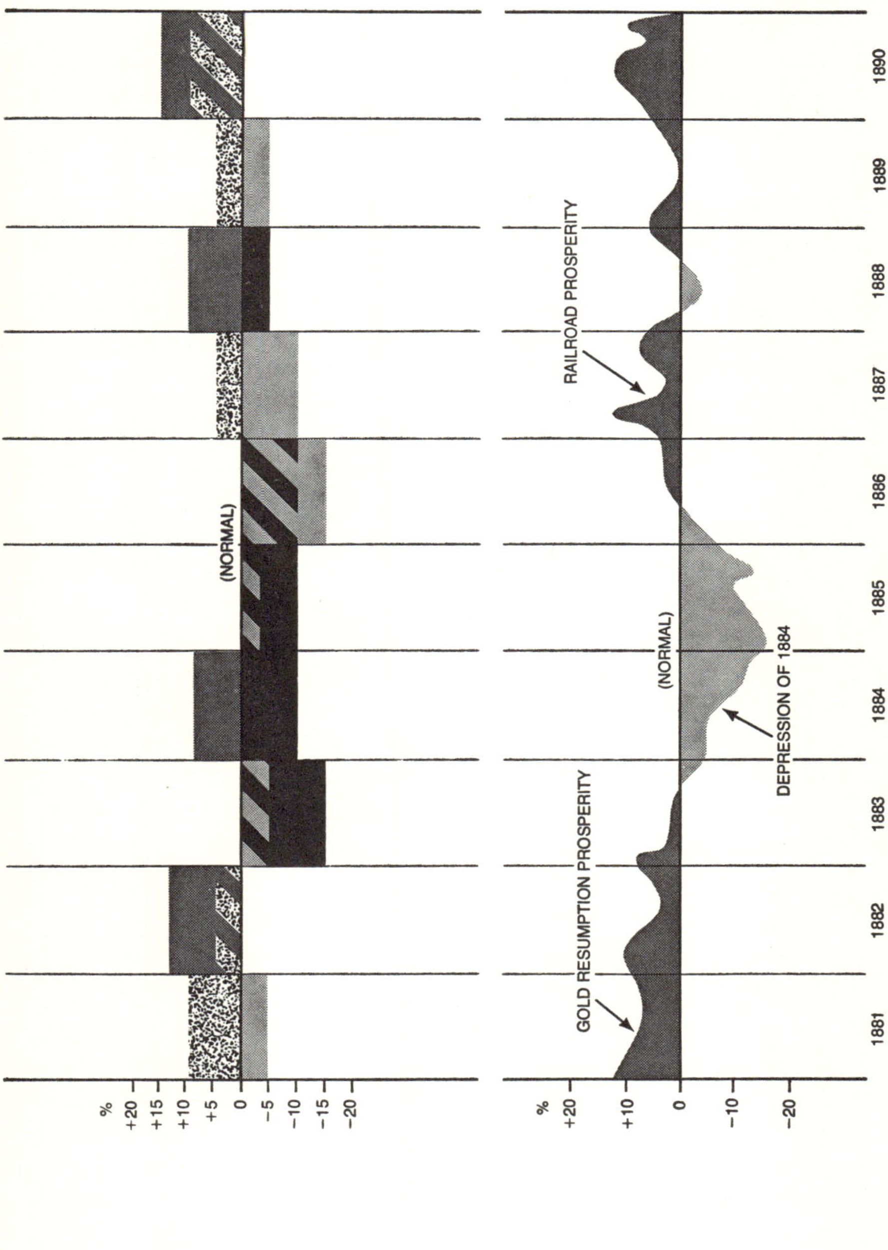

%
+20
+15
+10
+5
0
-5
-10
-15
-20
(NORMAL)
%
+20
+10
0
-10
-20
GOLD RESUMPTION PROSPERITY
DEPRESSION OF 1884
(NORMAL)
RAILROAD PROSPERITY
1881
1882
1883
1884
1885
1886
1887
1888
1889
1890

cycle; i.e., on the transition back to the cold side, as rainfall picked up again.

A period of prosperity matured during wet 1880 and lasted through most of 1883. Unsteadiness developed during dry 1881. Another peak in the boom occurred during wet 1882. The decline that followed was associated with dry 1883. While the depression of the middle 1880s centered in a dry period, wet 1884 failed to show any effect. The depression kept getting worse. Perhaps this was because it was dry in the eastern part of the United States, including New York. Generally speaking, however, prosperity—or the lack of it—is an integral, holistic condition pertaining to a country as a whole. Prosperity does not appear to be generated or caused by conditions in any one isolated section.

The recovery in 1886, when it was dry, was an unusual event. For a brief time, the business trends ran counter to the weather trends, demonstrating that, at times, economic and other non-meteorological factors may become more powerful than the weather trends.

This cold-dry decade was worldwide, as was the warming trend at its closing. As 1889 approached on the sunspot cycle, it began to warm up again. As is very often the case, rainfall revived along with temperature, for in the long run the two variables fluctuate together in a parallel fashion—except when it is temporarily warm and dry or cold and wet.

1891-1900

The 1890s saw the occurrence of the last part of the long cold-dry phase of the 100-year cycle, with the bottom being reached in 1895. The drop in temperature in the early 1890s was worldwide, as was the tendency toward higher temperatures in the middle of the decade.

The fifth depression era occurred in the 1890s during the end stage of the cold-dry phase of the 100-year cycle. (This cold phase had been interrupted twice, once in the 1850s and again in the 1870s, by reversals of temperature to the warm side, with the in-

Figure 26 1891—1900

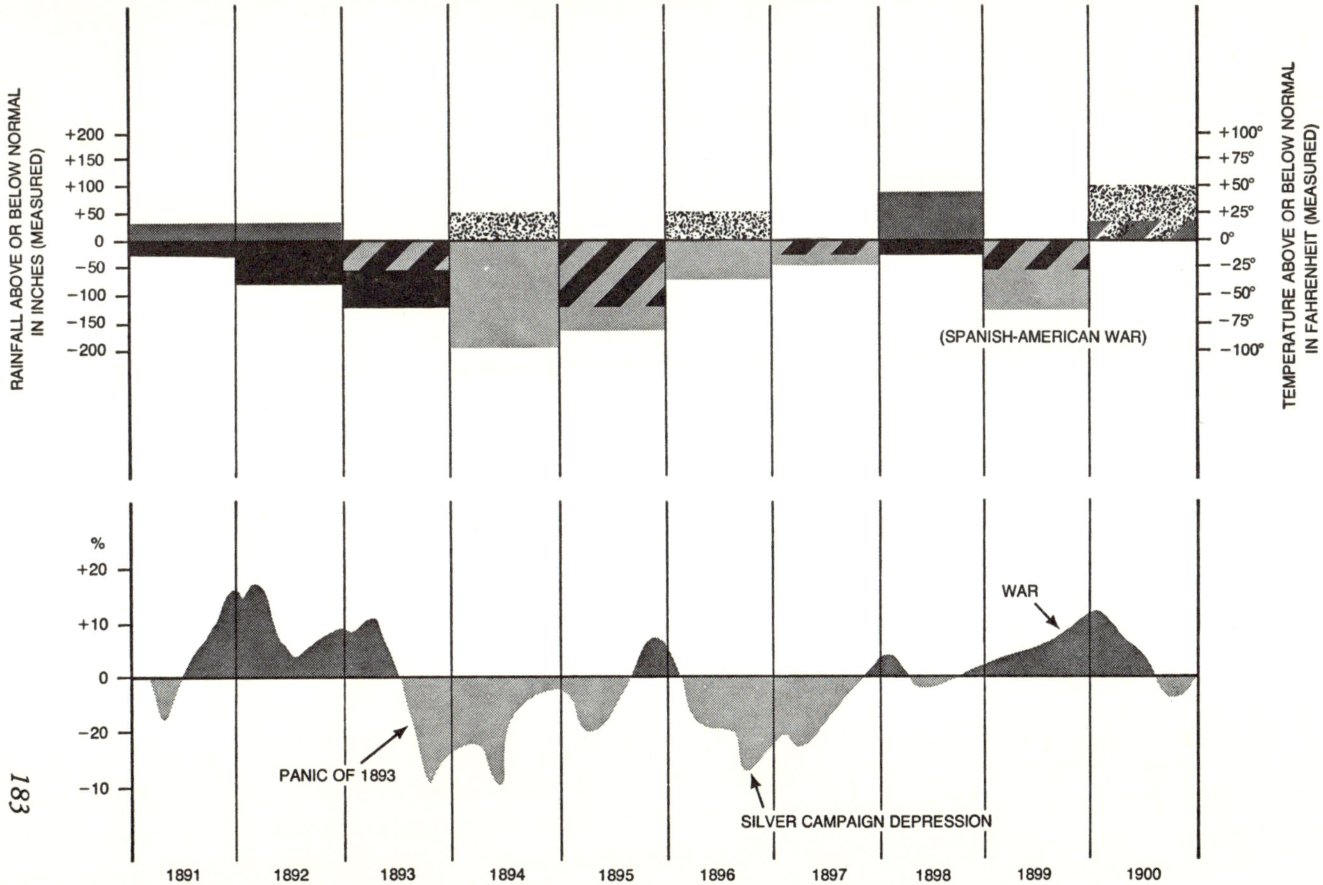

variable recoveries in rainfall, and booms, on the transitions back and forth.) This was a major era. The depression occurred in the center of a long period that had averaged dry, and the periods of prosperity occurred at both ends of the decade where the weather trends averaged wet. The boom of the late eighties continued as long as it remained wet, then it ended in a dry year. The temporary recovery in 1895 was mostly conditioned by economic, as opposed to weather, factors. The recovery associated with wet 1898 kept on during dry 1899 (Spanish American War). Warmer temperatures on the approach to the twentieth century were also worldwide. Cold 1899, wet 1898, and warm 1900 were conspicuous throughout the planet. Perhaps the drop in 1900 was a reaction to the end of the war.

15

1901-1975—From Warm-Wet Booms to Cold-Dry Depressions

1901-1910

The seventh boom era occurred on a major climatic transition from the cold side of the 100-year cycle to the warm side. This took place from 1898 to the 1920s, and lasted, as such major boom eras generally do, as long as the warm-wet phase of the 100-year cycle lasted.

The warm phase of the 100-year cycle really began at 1900; but beginning with 1902, it suffered an interruption before it had developed very far. There was a tendency toward a depression era from 1908 to 1911, in another secondary cycle, when dryness and coolness prevailed.

1911-1920

The weather trend temporarily became warmer and wetter after 1912. Again, after 1913, it suffered another interruption. Because

Figure 27 1901—1910

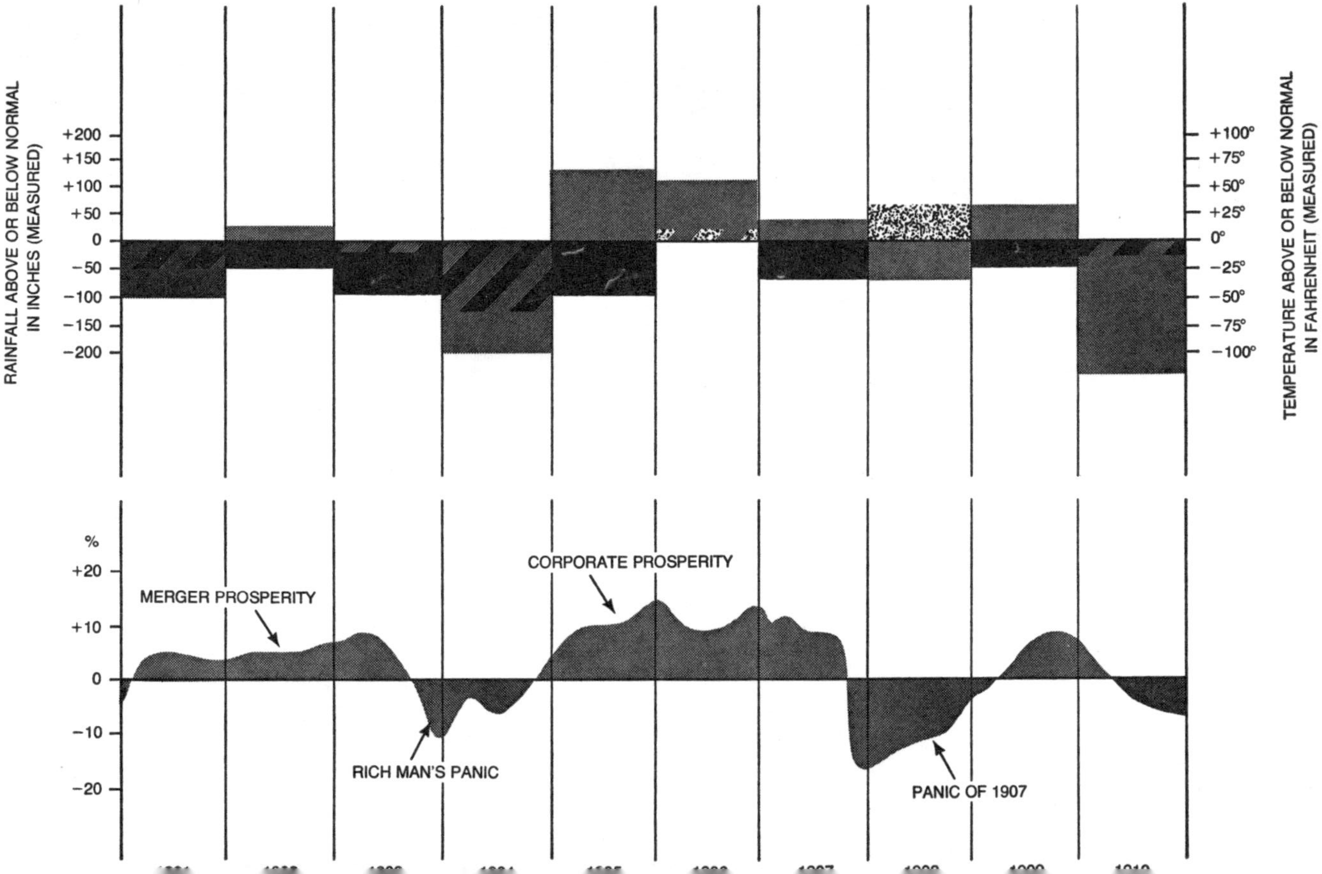

gure 28 1911—1920

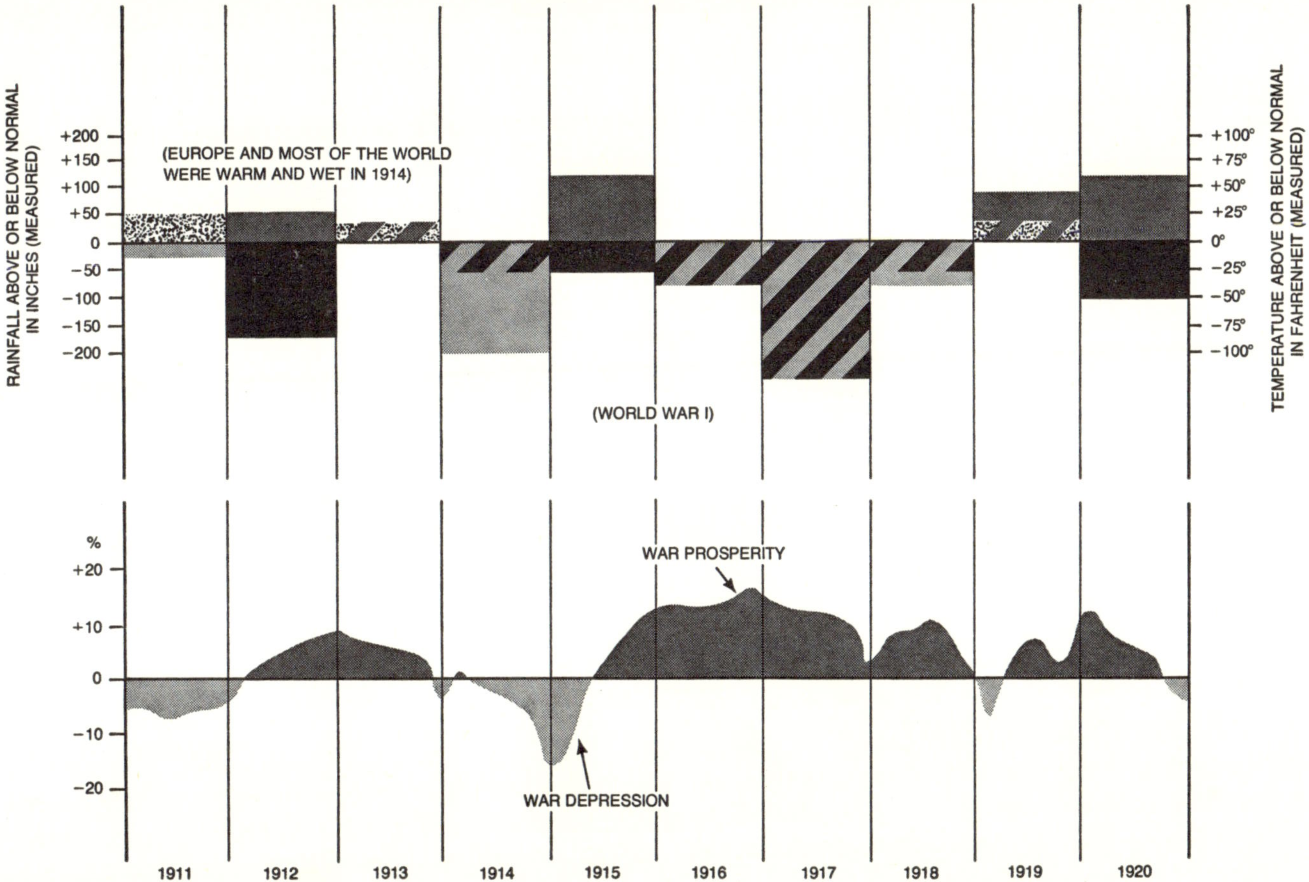

1914 was dry in the United States, there presumably would have been a recession that year even without a war scare. War production maintained high volume through three dry years, during which there is little doubt that a depression would have occurred had there not been a war at the time. As it was, production declined across the dry years and rose as rainfall picked up.

The cold years in the second half of this decade brought with them many outstanding democratic reforms, among them woman suffrage on a federal scale. Some twenty new democracies were formed at the end of World War I. The sharp drop in temperature spurred on the Bolshevik Revolution, and the continued cold helped bring on the revolution in Germany.

But none of these dramatic upheavals resulted in permanent democracies. The reason for the failure of these democratic revolutions lies in the fact that it turned warm-wet, then warm-dry. Such a sequence of weather trends has never failed to produce totalitarianism in some form or other. With a long warm period ahead, democracy had little chance to take root.

1921-1930

For the United States as a whole, 1921 was the hottest year on record, followed by the so-called Coolidge era of prosperity—which had nothing at all to do with Silent Cal. Six out of eight consecutive years were wet, and no such sequence at any time in history has failed to produce an era of prosperity. This was the last, and the strongest, warm-wet trend of the warm-wet phase of the 100-year cycle.

Economic factors had been strong enough, apparently, to bring about a revival during dry 1924, when it was also cold. The momentum of the revival had been vigorous enough to carry through dry 1925.

Note, however, a fairly strong tendency for the volume of business to have dipped during that year. Evidently, the drop at the end of 1927 was wholly economic in character. Wet 1923, 1926, and 1929 were the highest and broadest peaks in the boom.

Figure 25 1921—1930

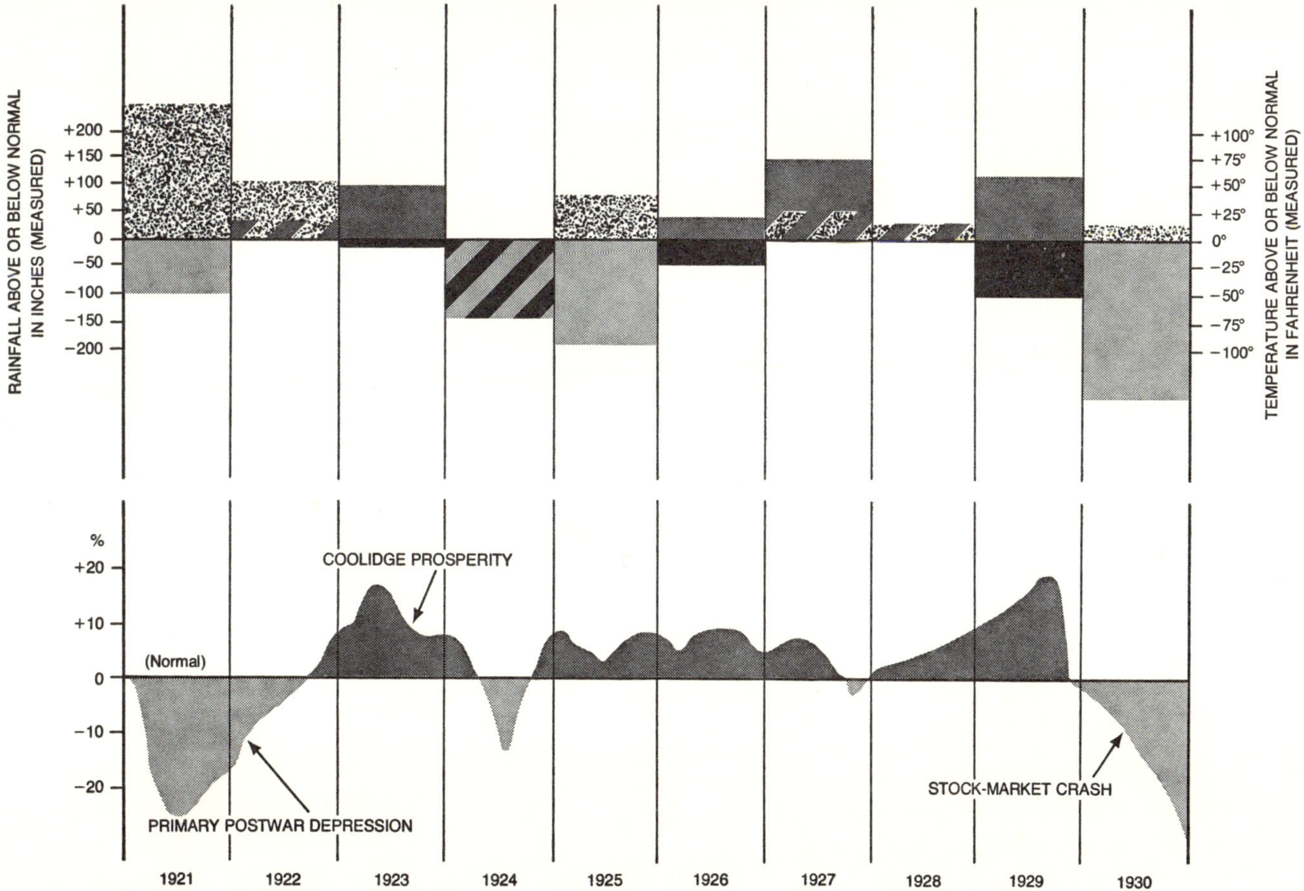

From 1898 to 1929, prosperity, in general, ruled over depressions. It was nation-building time again, but also a time for world wars as well as for Golden Ages.

Twenty-three out of the thirty-one years of this period were prosperous, and eighteen out of the thirty-one were wet. The depressions were short and the booms were long. Three dry years during World War I were boom years because of the war.

But the year 1930 was the driest over the country, and it was also the year following the great crash, the most severe in the history of the United States.

1931-1940

The sixth and last depression era, a major one, repeated the pattern of the 1820s. This was in the 1930s when again there occurred the warm-dry phase of the 100-year cycle. The intervals between the major warm-dry phases of history are by no means constant, but they average 102.5 years from the fifth century A.D. to the present time. The main warm-dry periods have centered on the following dates: 547, 613, 710, 817, 920, 1025, 1130, 1270, 1365, 1455, 1530, 1630, 1690, 1825, and 1935.

Each of these phases has been followed by a shift from the warm to the cold side of the 100-year cycle.

Each has been followed by eras of very evident commercial and business stagnation after a temporary period of prosperity had intervened during the cold-wet phase of the cycle.

Each warm-dry period has been associated with a major economic breakdown, marked by unemployment, suffering, low production, and often state charity of some kind. Each warm-dry phase has brought about a climax of despotism, dictators, and New Deals. Diocletian, Tamerlane, Peter the Cruel, Domitian are some typical hot-drought specimens of the past. In this century we were provided with Hitler, Mussolini, and Stalin.

During cooler and wetter 1932, there was a tendency for the decline in business volume to reverse and begin to climb. There was enough momentum to carry itself partway through dry 1933, then

Figure 30 1931—1940

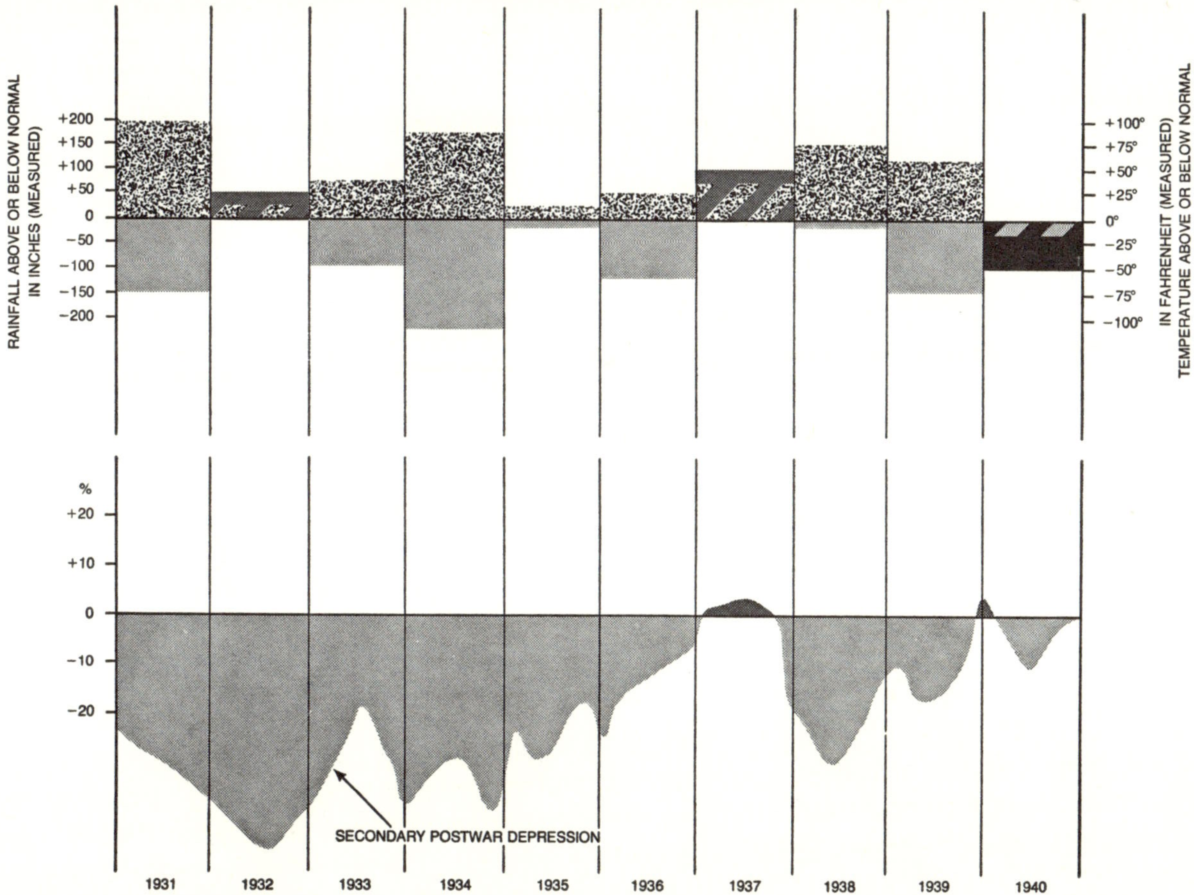

decline set in once again. After 1934, the rainfall trend was upward. The wet year 1937 brought mild recovery in business.

1941-1950

Under ordinary circumstances, there probably would not have been a revival to the boom side in 1941, but the war lifted business prosperity and prevented a decline during dry 1943. The drop in 1946, of course, was due to conversion to peacetime economy.

Notice that three out of the nine years were cold. This is the first showing of the downward temperature trend, which has now overtaken our country and the world as a whole.

1950-1975

The eighth boom era, in the 1950s, occurred on the reverse side of the 100-year cycle, as the warm phase was changing to the cold and dry throughout the 1950s and the 1960s. There have been boom of the 1950s was a feature of the cold-wet phase of the 100-year cycle.

In terms of eras, therefore, there have been no exceptions since 1794 to the rule that boom eras average wet. It has been generally cold and dry throughout the 1950s and the 1960s. There have been slight recoveries in rainfall and in the economy, and serious depressions have been avoided, much in the manner of the 1830s and the 1840s (see Figures 5 and 6). The occasional brief peaks of prosperity were due to the various "police actions" in Asia.

Droughts have been getting worse over much of the world; and with the increasing dryness, temperatures over Earth have, on the average, been gradually dropping. The little peak in rainfall which occurred in 1952 matured because that year lay at the crest of the seven-year rainfall cycle, which hardly ever fails to show an increase in rainfall. The recovery in rainfall for 1958 and 1959 belonged to the same cycle. The drop in temperature at that time was also part of a common cycle.

The situation in the 1830s and the 1840s, following the

Figure 31 1941—1950

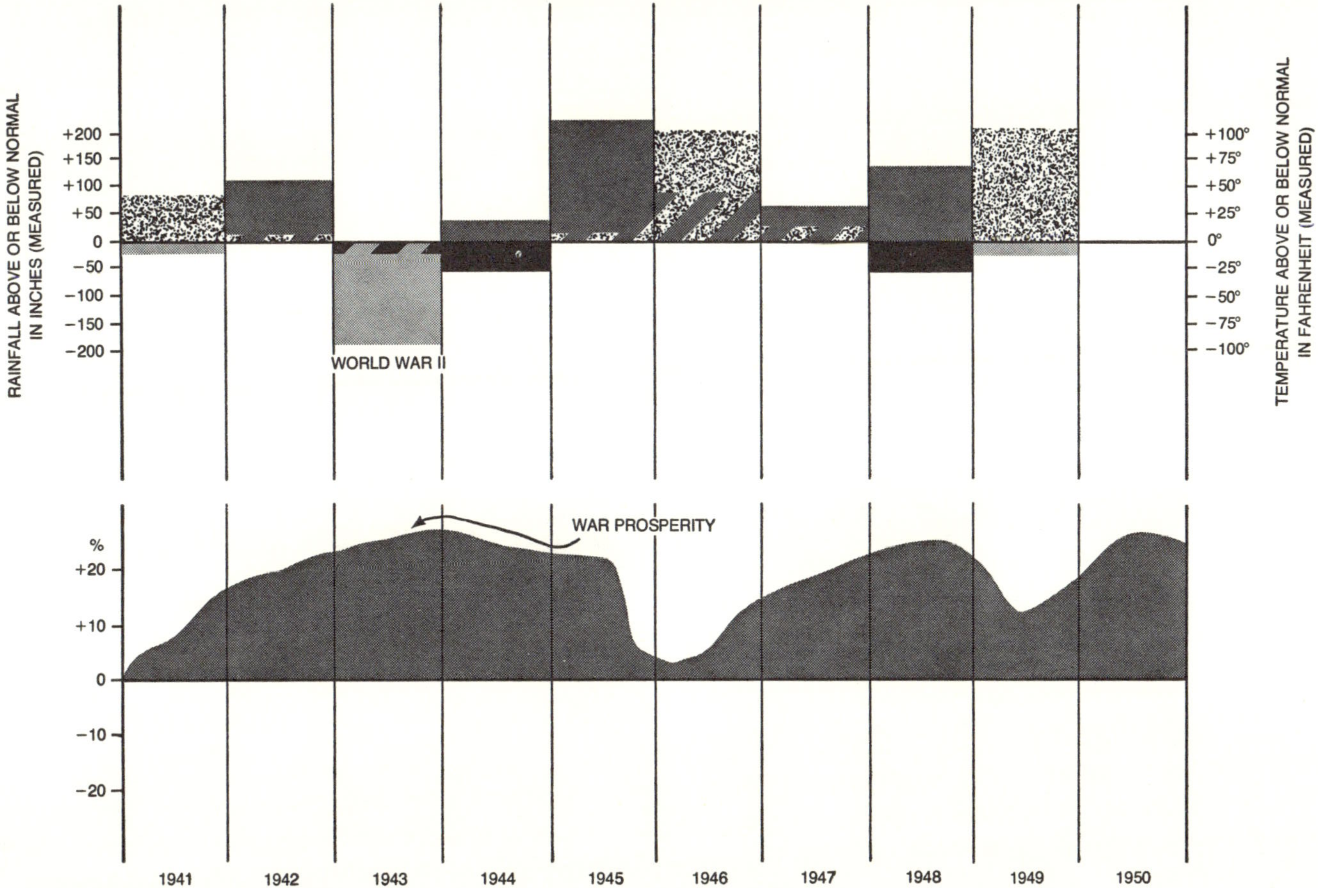

Figure 32 1950—1975

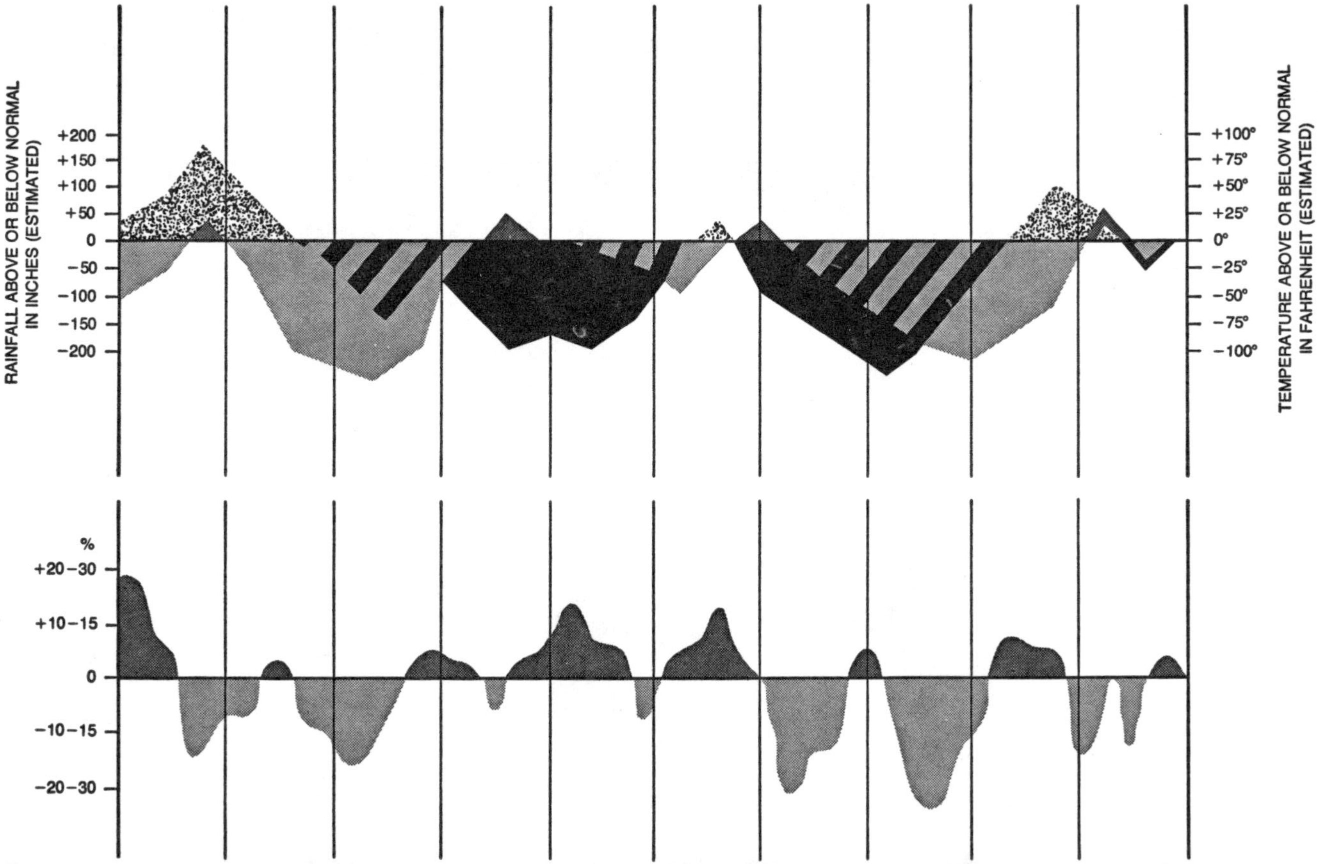

warm-dry phase of the 100-year cycle in the 1820s, repeated itself in the 1960s.

The 100-year cycle in rainfall that has been running uninterruptedly down through history, since the time of ancient Greece, points with convincing force to a long dearth of rainfall through the 1970s.

The 1970s should correspond to the 1850s. The drought period will turn from cold to a mixture of cold and warm, as in the 1860s. Drought cycles in the tree-ring curves of the western trees predict this severe dry era.

This is of importance to the businessman from a practical standpoint, for it opens up the possibility of long-range business forecasting based on long-range weather forecasting. If a pragmatic businessman studies these charts and concludes that the cold-dry period which we have just entered is about to bring us some bad times, he is most probably correct. How he utilizes this forewarning may indeed determine his survival or his failure in the next decade.

The two major wet and two major dry periods of the 100-year cycle were traced back to the sixth century B.C. in four chronologies. From the types of rulers and types of events prevailing in these four periods it was evident, without specific economic documents, that the major "boom" and "depression" eras can be identified throughout history. Human behavior has been similar in pattern in each of the repeating phases, and in each case the pattern resembles that of the last two cycles. There are two places on this cycle when prosperous times can be expected with almost absolute certainty, and two places where depressions can be fully expected.

The prosperous times occur on the transitions from cold to warm and warm to cold, and into the warm-wet and cold-wet phases of the cycle. In other words, eras of prosperity occur during wet periods, and, regularly, only then. Eras of prosperity are of two kinds: One is associated with nation-building, imperialism, industrial revolutions, mergers, the growth of monopoly and a trend toward statism and socialism. The other is associated with nation-falling and another outbreak of imperialism—but with insurrections, democratic reforms, and a trend back toward individualism.

Both eras are filled with underlying waves of optimism and assertive aggressiveness. The first is optimism in regard to national destiny; the second in regard to the destiny of the individual. Since 1794, booms have occurred on all the transitions from cold to warm and warm to cold and into the wet phases of the cycles and regularly at no other times.

In terms of major eras, depressions have without exception been dry since 1794. Depressions have occurred during most of the warm-dry and cold-dry periods since 1794 at temperature maxima and minima and regularly at no other times. Eras of depression are of two types:

The first is a breakdown of the economic system at the end of a nation-building period, when centralizing trends reach a climax economically and politically. This depression is characterized by state paternalism, oppression, and socialistic-communistic trends.

The other type of depression is associated with the collapse of individualisitic economy, periods of inadequate government, and trends toward anarchy and chaos.

The first type is warm; the second is cold.

Each of the four eras, two prosperous and two "unprosperous," occurs repetitively in the 100-year climatic cycle. Remember, then, this sequence:

1. Hot-drought depression, at the end of a long warm period.
2. Cold-wet boom at nation-falling time, at the opening of a long cold period (where we have just been on the cycle).
3. Cold-dry depression in the midst of the cold period, when rainfall stops or declines (where we have just entered on the cycle).

Business trends vary consistently with the weather trends. Prosperity ordinarily lasts as long as the rainy periods last and halts when it stops raining. Depressions last as long as the dry periods, and end when it starts raining again. Time after time, single wet years will be associated with single boom years, and single dry years with single depression years. The next era, to begin shortly, is an era of depressions.

PART III

THE NEXT 25 YEARS—AND HOW TO LIVE THROUGH THEM

16

The Grim Specter of World Famine and Financial Depression

Those two bold cartographers who set down a Roadmap of Time gained little recognition for their efforts during their lifetimes. But Raymond Wheeler and Selby Maxwell would have been pleased to observe that their work will be disseminated to the public at a time when, more than any other crisis in the present epoch, mankind desperately needs such guidance.

As I write this early in 1975, experts are predicting bankruptcy, social breakdown, and starvation for as many as one billion people within the next twelve months. Some thirty nations, mostly in Africa, South Asia, and the Central American-Caribbean area, will be among the first to be affected.

In a two-part feature for *Science News,* John H. Douglas warns that broad-based food reserves and new technological advances are needed to stave off worldwide famine. Douglas reports grimly that global food reserves are at their lowest level in two decades and that "with malevolent coincidence" the world's climate has begun to change. Douglas quotes weather scientists who say that the Earth's temperature has lowered 2.7° F. since 1945. Weather over the last half century has been the warmest in a thousand years, and a cycle of colder temperatures can be expected. There will be a major shift of rain patterns and deserts, and northern latitudes will be faced with a shorter growing season—just when a starving world will be looking for more food from their fields.

According to a study conducted under the aegis of the United Nations, more than 700 million in less-developed countries are suffering from malnutrition. All experts state unequivocally that the leadership of the United States will be essential to check world famine and to sustain starving millions with its agricultural wealth.

But in eight years, 1965 to 1972, the United States provided 84 percent of all food aid contributed worldwide by industrial nations, according to figures compiled by the U.S. Agriculture Department. Now, with grocery bills climbing 36 percent in 1973 and 1974—largely because of increasingly tight food supplies—both citizens and politicians are beginning to feel a bit less generous toward the stock in our own dwindling larder. Resentment has also begun to grow against such a nation as India, the single largest recipient of past food aid, which fails to show concern for population control and which recently directed vital funds away from food productivity into the development of its first nuclear weapon.

J. Merrill Anderson, Iowa Farm Bureau leader, told farmers assembled for a conference in Des Moines that it was time for the United States to establish sound practices in its national and international policies before the greatest food-producing section of the world was inadvertently destroyed.

"Some well-intentioned people are advocating either elimination or substantial reduction in animal agriculture," Anderson remarked. "If we were to follow these suggestions, we would not get any more food to the starving. We would bankrupt grain farmers and then produce less.

"However, we cannot ignore world suffering. The United States has pledged temporary food aid. We can give more technological and other types of assistance if [underprivileged nations] decide they will do what they must do to help themselves."

Anderson, who in November, 1974, attended the world food conference in Rome, observed that official policies must have been wrong in the past; for although the United States has done more than any other nation to assist other countries, he heard nothing but vicious criticism of the United States. Anderson suggested consideration of a national policy of tough-minded humanism or tough-minded compassion:

"We will help these desperate people only if steps are taken to make lasting progress. It appears highly probable that giving food without any requirements will only result in more suffering in the long run."

Dr. Garrett Hardin, professor of human ecology at the University of California in Santa Barbara, suggests that the United States adopt a "lifeboat ethic." It is not possible to share wealth, Dr. Hardin points out, only poverty.

"So long as nations multiply at different rates, survival requires that we adopt the ethics of the lifeboat," Dr. Hardin told Ronald Kotulak for *Chicago Tribune Magazine.* "A lifeboat can only hold so many people. There are more than two billion wretched people in the world—ten times as many as the population of the U.S. It is literally beyond our ability to save them all. . . . Both international granaries and lax immigration policies must be rejected if we are to save something for our grandchildren."

Dr. J. George Harrar, president emeritus of the Rockefeller Foundation and principal architect of the "green revolution" that developed hybrid grains exceptionally responsive to fertilizers, foresees massive world famine if the United States should suffer one or two major crop failures. "Climatologists anticipate that, based on the cyclic nature of the weather, North America could well have a severe drought in the near future. I'm scared. We could have a lot of starvation by 1980. It is unrealistic to think this could not happen."

Financial experts are predicting a market crash that will make

1929 seem mild by comparison. Unrestrained inflation is forcing men and women to spend so much on the necessities that luxury and leisure items are becoming dregs on the market. Increasing scarcity of raw materials and inexpensive food will keep prices arching upward and continue to fuel cost-push inflation. There is a terrible economic crunch coming, these financial pessimists warn, and everyone is going to be affected.

The December 16, 1974, issue of *U. S. News & World Report* presented a composite view of the 310 securities executives who responded to their questionnaire survey. The consensus of the respondents was that "the U.S. economy is going to pull out of recession rather than sink helplessly into depression," but no doubt "the economic life of most individuals and businesses is sure to become decidedly rougher before any turn comes for the better."

The survey indicated that the overwhelming majority of the securities executives felt that, with downward trends gathering momentum, there is "almost no chance for a strong recovery in the stock market in the next few months. Declining profits, competition from still-high bonds yields, and no more than modest gains against inflation are expected to sap the vigor of any incipient bull market in stocks for a time."

"And I suppose," I told John Cejka, "the real purpose of all these experts' studies of current trends is to enable man to make sensible predictions about his future."

"Of course," Cejka nodded, "but it would take many years of careful study with standard mathematical techniques before any predictions or extrapolations can possibly be raised from more than a purely speculative level. Because we have so many clues from which to work, we must be certain we have evaluated each of them properly.

"For example, in about 1926, A.L. Tchijewsky made a considered prediction that the years 1927 to 1929 would be a period of great human activity of the highest historical importance, a time that would change the political chart of the world. All this, he said, was the result of a similar sunspot maximum in 1879. Tchijewsky further assembled this prediction on the basis of a sixty-year and a thirty-year sunspot period.

"Well, as we now know, nothing of much importance occurred during this maximum except for a terrible economic collapse; and it is obvious that Tchijewsky had nothing of such a negative nature in mind. The collapse may certainly have had something to do with the political chart of the world as well, but hardly of the caliber of the highest historical importance.

"Maxwell and Wheeler agreed with certain of the points in Tchijewsky's extensive investigations, namely that cultural activity does vary with the sunspot cycle. Certain multiples of the sunspot period seem very important both in the rhythms of history and in climatic fluctuations. They agreed that sunspot minima are likely to be relatively quiet times—or at least to be near quiet times.

"On the other hand, they disagree with Tchijewsky's contention that sunspot maxima coincide regularly with maxima in human activity.

"For a period of time, several of these intervals will repeat themselves, then vanish, only to reappear at a later time. The problem is quite complex. You see, quite as good a case could be made for the claim that sunspot maxima are also quiet in most parts of the world. Tchijewsky had made the universal error of oversimplifying the problem—both as to fact and as to theory."

"And you feel that the Maxwell-Wheeler weather-energy research overcomes those shortcomings?" I asked him.

"Brad," he said, pointing a pistol finger at me, "you know that I am only impressed by track record. I've tested the Maxwell-Wheeler research extensively against the so-called orthodox weathermen and their forecast methods, and Maxwell-Wheeler is vastly superior. Howard Baker offered an excellent service based on the work of these men until the early 1950s, and now we have been able to extend these services through our Cyclomatic Engineering Corporation.

"And nations can plan accordingly," I said. "but with caution."

In 1975 we are facing difficult times in weather trends in the cycles approaching. There will be dramatic changes in world conditions—political, cultural, and economic—which are of great importance to every one of us.

In Dr. Wheeler's opinion the correspondence between the mathematical and the actual curves was close enough to justify the expectation of certain trends in the future.

Again, Dr. Wheeler's intent in charting such curves was to present the following urgent facts:

- World temperatures are falling.
- We are now in the cold phase of a 100-year cycle. The remainder of this century will be spent on the cold side of the longtime average.
- Accompanying the drop in temperature will be long, severe droughts, from which there will be no major relief until about the mid-1980s.
- After the cold phase has passed, world temperatures will probably begin to climb. The resumption of warmer weather will be accompanied by a great abundance of rainfall, after which there will perhaps be a recurrence of droughts such as those of the 1930s.
- At that time, there may occur the heat climax of a 1,000-year cycle (the previous heat climax occurred around A.D. 1000). After the heat climax, which will peak in the 2040s, it may be that the northern lands will begin freezing up again.

We are now entering a cold-dry period. But as John Cejka observes, "It is during the warm-dry phases of the 100-year cycle that civilizations all over the world become sick, and we are just leaving such a period."

"If this is a recurrent situation on the 100-year cycle, why are you talking about an old world dying and a new one being born? This phase should be relatively transitory."

"You must remember the 500-year climatic rhythm when the periodic illnesses of a society are intensified. World civilizations manage to get in really serious trouble toward the end of these 500-year rhythms, and we are in one of those times now, too."

In Wheeler's assessment of the recurring cycles, every other 500-year period in history belongs to Western civilization. Every intervening 500-year period belongs to Eastern civilization. Dr. Wheeler found that at the end of every other 500-year cycle, major revolutions occur over the entire world and result in drastic reorganizations of society. According to his projections, we are in a revolution of this sort now, comparable to the collapse of medieval culture and the beginning of the modern world.

"The Maxwell-Wheeler research," John Cejka said, "indicates that societies all over the world become seriously ill at least once every 100 years, desperately so every 500 years. Every 500 years, the 100-year sickness rhythm and the 500-year sickness rhythm converge. That's what is happening right now."

"What are some of the symptoms of such a convergence?"

"One is an unusually severe depression. Another is the moral tone of society—especially its manner of rearing its children. There are hundreds of these symptoms: the kind of art prevailing at the time, the emergence of state socialism, decadence in the government."

Now in the 1970s, the 500-year era which has belonged to Europe and her offspring, North America, is coming to an end. The next 500 years of history will belong to Asia.

"It seems highly certain," Dr. Raymond Wheeler wrote, "that the initiative is again passing from the West to the East for a 500-year period, as it did 1,000 years ago and 2,000 years ago. All of Asia is waking from a sleep of 500 years. Needless to say, the future of America is tied into this great convulsion and must be appraised in terms of it, or not at all.

"The present 500-year cycle is due to end around 1980. The changes that are now taking place will, of necessity, alter many of our patterns of behavior. Our economic system and the world of business are not exempt."

It was Dr. Wheeler's contention that one should always seek to obtain a general view of things so that the more specific and concrete happenings may be fully understood. "The latter derive their meaning from the former," he observed. "The parts derive their meaning from the whole."

In order to better understand the events of today and tomor-

Figure 33 THEORETICAL TEMPERATURE AND RAINFALL CURVES BASED ON PAST RECORDS

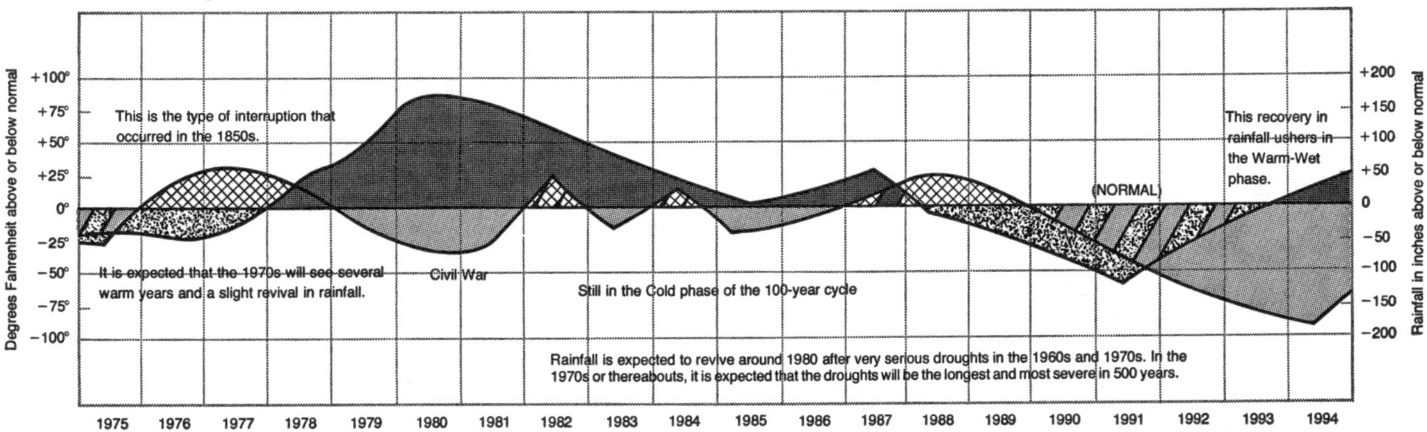

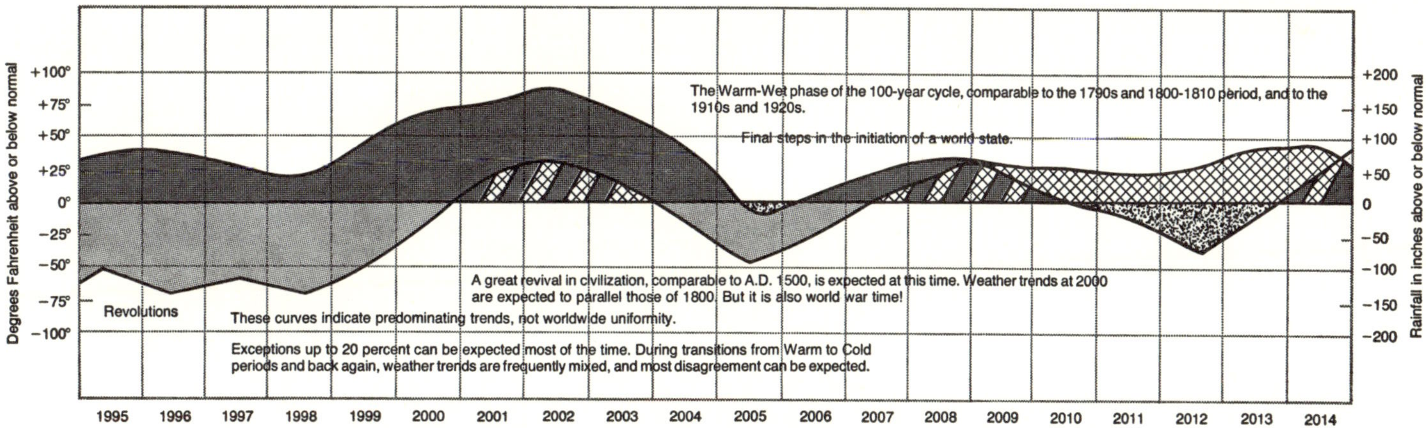
Degrees Fahrenheit above or below normal
+100°
+75°
+50°
+25°
0°
−25°
−50°
−75°
−100°
Rainfall in inches above or below normal
+200
+150
+100
+50
0
−50
−100
−150
−200
The Warm-Wet phase of the 100-year cycle, comparable to the 1790s and 1800-1810 period, and to the 1910s and 1920s.
Final steps in the initiation of a world state.
A great revival in civilization, comparable to A.D. 1500, is expected at this time. Weather trends at 2000 are expected to parallel those of 1800. But it is also world war time!
Revolutions
These curves indicate predominating trends, not worldwide uniformity.
Exceptions up to 20 percent can be expected most of the time. During transitions from Warm to Cold periods and back again, weather trends are frequently mixed, and most disagreement can be expected.
1995
1996
1997
1998
1999
2000
2001
2002
2003
2004
2005
2006
2007
2008
2009
2010
2011
2012
2013
2014

row, Wheeler advised, one should study occurrences at the termination of previous 500-year cycles. "While specific events do not repeat themselves in history, *types* of events unquestionably do," he commented. All of the signs point to a continued repetition of this rhythm—which means that the rest of this century is due to be severely cold. We are now into the cold phase of the 100-year cycle and the fifth and severest cold phase of the 500-year cycle.

Today, once again, as another 500-year cycle is drawing to a close, outworn institutions and outmoded patterns of thought are no longer adequate. Once more we are heading into a cold period that marks the end of a cycle in our historical epoch. The twentieth century may be a crucial period in the history of civilization. In the coming centuries, climatic changes will become more severe, the warm and cold phases longer. This will mean more violent transitions from one side to the other, with an increase in storminess, floods, and, doubtless, an increase in events such as earthquakes and volcanic eruptions. This will also mean longer warm-wet periods with an increased danger of tropical diseases, longer droughts with a greatly lowered human vitality, and a greatly reduced birthrate. In addition, unless political structures are stabilized meanwhile, there should develop intolerable despots during late warm times, and a degeneration into piracy and anarchy during cold times. Such effects could produce undreamed-of slaughter and costs during unimaginably horrible world wars during the climatic transitions. This would also mean devastating economic depressions with unheard-of suffering and want.

But on an even larger scale, if his interpretation of the data is correct—and if the longtime process of acceleration has reached its limit or nearly so—Wheeler feels that the only conceivable expectancy is that of a reversal to a long series of decelerating cycles during which climate will deviate more and more from optimum conditions. Eventually, the decelerating process would reach the point at which the temperate zone would become tropical or semitropical on one side of the axis and severely cold, perhaps cold enough for ice ages, on the other side. While the suspicion that we are presently between two ice ages is common among geologists, the implications of this possibility have never been seriously faced.

"It would seem more than a matter of wild speculation that this assumed reversal of climatic cycles may be upon the human race within the twentieth century," Wheeler declares.

If we are about to undergo the fourth cyclic reversal and if the first three reversals introduced new life forms, what dramatic evolutionary changes might mankind be about to witness?

Wheeler says that this cyclic reversal, falling as it does during the present cold-dry period, will be "difficult to live through." If the history of climate as outlined by Wheeler should repeat itself, he warns that the "transitions will be accompanied by an increase in the severity of the earthquakes and the vulcanism that attended the lowering of the continental altitudes, with marine invasions on the upswings and mountain building on the downswings."

In other words, Raymond Wheeler's portrait of the last twenty-five years of our century is very much like that revealed in the "Earth Changes" material of the psychic Edgar Cayce, the American Indian prophecies of a Time of Great Cleansing, and the interpretation many fundamentalist Christians place on apocalyptic passages in the Book of Revelation.

In my book *Mysteries of Time and Space* (Prentice-Hall, 1974), I state my belief that these Judgment Day revelations are largely misinterpretations of visions concerning a time of transcendence and transformation for our entire species. I quote Eugene A. Albright who, in his *Uni-Chotometrics,* may have glimpsed what dramatic changes mankind might be about to make as he undergoes cyclic reversal:

"The incapacity of the human organism at this point to directly convert matter into its own structure is one of the factors being changed in nature. The next evolvement, or the next evolving technique of the human organism, will be the opening up of two specific functions which, up to this stage of development, have been latent.

"One of these is the capacity to control the environment completely; to cause matter in the environment to disintegrate and restructuralize directly on an energy level. The other is to structure the function of the body and replenish it without necessity for food, either plant or animal life."

Albright, too, states that there have been periods in the Earth's past when the vibratory rate was stepped up. "Many who cannot adjust to higher rates of vibration are destroyed. This may appear cruel, but nature and that which is natural has always prevailed. . ."

Just as vertebrate life patterns led to mammalian forms and they, in turn, to our immediate ancestors, Albright sees this new infusion of energy as causing many structural changes in the human organism itself. Nature, all plant and animal life, will change as well.

"Within a relatively short time there will be nothing . . . recognizable, using our current knowledge of physics and matter, chemistry and biology," Albright writes, adding that man must become totally related to the new environment.

"To us who are living in this era of change, it is important to understand that we are a part of one universal uni-polar magnetic field," Albright advises. "This universal uni-polar magnetic field is a process of implosion, gradually increasing. . . . This is the direction life has always taken; it has increased in time. . . . the evolving circumstance has been increased every year at a given time; a new influx of energy is infused into the field related to the human being."

17

Future Politics and Economy During a Cold-Dry Period

Whatever changes may be in the process of altering the very roots of our civilization—perhaps the very basics of our human organism itself—Dr. Wheeler's own notes may provide us with a sensible guideline for surviving the next twenty-five years. Based on his research which stressed the comparing and the contrasting of past cycles of weather energy and their effect upon all human affairs, Dr. Raymond Wheeler made the following predictions for Cycle 27, which we have now entered.

*There will be a prolonged period of civil strife, class struggle, and civil wars all over the world, as trends shift to democracy. There will be temporary despotisms. "Another important factor can be expected to complicate the crime picture and perhaps increase the extent of crime in the decades ahead. This factor is the increasing cultural instability of nations and peoples the world over.

"The most important large-scale weakness to be expected is . . . in government. We can expect to pass through a period when governments will be weak. Disregard for law will reach a longtime high. Local governments will resist the central government sitting over them. Quarreling between factions will increase.

"Racial and religious minorities will demand more privileges of the state. The problem of states' rights will become a burning issue. Decentralization of the federal government will be demanded. Disorder will increase in the form of uncontrolled fraud and graft, rackets, and swindling. Piracy will become a problem in some parts of the world. There will be a revival of secret orders and societies, some of them bent on taking the law into their own hands. Mob violence, race riots, and the like will be more frequent. Strikes will increase as labor becomes more and more demanding." All official forms of race and religious discrimination will disappear in the United States and the other democratic countries.

*Unless they make drastic democratic adjustments to a rapidly changing world, the present ruling classes and parties of the world will disappear.

"It is obvious, of course, that the more totalitarian states of today will most likely to be slower to give in to popular demand than the more democratic nations," Dr. Wheeler wrote. "From general knowledge, it is obvious also that many of the demands for more participation in the economic and political life of the state or the group which will be made by the working classes will be called socialistic by those whose interests are at stake. Remember that woman suffrage, not a socialistic measure at all, but a profound democratic measure, was branded by its opponents as evil, because it was 'socialism.' (The term socialism is used here, correctly, in the light of history, as describing that political organization which gives to the state or group those rights, privileges, controls, and ownerships that in a democratic society, during democratic times, have been demanded by the individual. Socialism has never existed historically—nor can it exist—without a delimitation of the franchise and a trend toward despotism.)"

If, then, it is going to turn cold, Wheeler tells us, the political changes which will occur all over the world will be in the direction of democracy, and away from socialism and communism. Those countries which have scrapped their democratic constitutions will have to revive them. Those countries which, already democratic, have centralized their political control and have limited local autonomies, will go through a decentralization process.

*The undemocratic countries will endure bloody civil wars. Specifically, Spain will suffer civil strife as another attempt is made to accomplish democratic reforms in that nation. There will be a succession of revolutions in the Latin American countries as more democratic forms of government and more democratic forms of economic control are being created.

All communistic governments will fall. Terrorism, mob violence, anarchy, will hold sway until reforms have been effected in those countries which have supported communistic regimes. There is a strong possibility that Soviet Communism will disintegrate because of a four-way civil war with North vs. South and East vs. West.

Both the warm and the cold periods of history have made their permanent and substantial contributions to the evolution of the modern state. Warm periods have furnished the concepts of cooperation, coordination, and integration. Cold periods have furnished the equally necessary concepts of individual initiative and freedom. Neither will permanently give way to the other. The result, of necessity, will be a compromise between the two in their extreme forms. The long-term outlook is a middle-of-the-road position. But the end of existing despotisms and international wars is near at hand, and the beginning of a period of civil wars and rebellions is imminent. Certainly the political behavior of the world, both qualitatively and quantitatively, harmonizes today with the supposed position on the climatic cycle, as it has harmonized with corresponding positions throughout history. How soon the wars will stop will depend in considerable measure upon the rate at which the temperature falls and upon a factor that cannot be measured—namely, the resistance that existing governments can offer to the inevitable rising tide of discontent and rebellion.

No one can tell precisely how long the expected cold period will last. Should temperatures fail to drop consistently during the next few years, the onset of rebellions may be delayed, and the international wars may continue; but not necessarily so. Climate is not the only cause of human affairs, albeit that it assuredly determines the main trends over a period of time.

If it should prove true that a major cold drought epoch is due during the latter part of the century, one's guess would be that the warm period would be short, perhaps not over a decade, and that by 1976 the long cold epoch would be well under way, with breaks to the warm side at, or following, the double minimum due in the middle or late 1980s and again in the 1990s. A long cold period would doubtless mean a more or less complete democratization of the world—a type of democratization that would not abandon those features of centralized government and those social responsibilities of the state that history has demonstrated to be workable and desirable. It ought to mean the last of dictators forever. It ought to mean that in the next world war—if one is fought—the democratic nations would win, or, if they should lose, then the less democratic nations will have won the war in vain, because they will of necessity have to change their ideologies soon thereafter.

*We must recognize that every alternate 500-year cycle has been characterized by the domination of Europe. The intervening 500-year cycles have been dominated by Asia. A 500-year cycle which belonged to Europe is now terminating. A study of historical cycles shows that a world convulsion is occurring second only to the emergence of rational thought in the sixth century B.C.; the fall of Rome and other ancient civilizations in the fifth century; the Renaissance after the collapse of the Middle Ages in the fifteenth century. The coming convulsion is comparable to the birth of Christianity in the first century and the birth of the modern nation as a feudal principality in the tenth. The next 500-year cycle will belong to Asia. The parts of the world that will develop the most politically and economically during the next 500-year cycle will be Oriental countries. The warmer countries and areas of the world, including the southern United States, will experience a dramatic awakening. We must learn to be more tolerant and understanding of other people in our own country and people in other nations.

The cold-dry phase of the 100-year cycle which we are about to enter is one of a long, or at least recurring, depression. The cold-dry phases have been the Dark Ages of history. Civilizations have disintegrated during these periods. Chaos, anarchy, and piracy have prevailed. The world has seethed with migrations. Orderly government and commerce have suffered through terrible times of civil war, assassins, robber barons, and militant societies. Can we be well enough prepared to alter the pattern so firmly established?

"Lest the reader be skeptical concerning the nearness of a serious cold-dry era," Dr. Wheeler wrote in the 1950s, "and lest he be led astray by the current stories about the world getting warmer, let him contemplate the following figures. The centers of cold-dry periods since the last days of the Roman Empire, as attested to not only by masses of miscellaneous reports of weather conditions but also by tree rings, have occurred as follows:

A.D. 460	1295
A.D. 560	1400
A.D. 650	1475
A.D. 745	1570
A.D. 845	1655
A.D. 955	1765
A.D. 1055	1865
A.D. 1175	

"Notice how these dates fell for the most part in the middle of the centuries. Variations occurred in the thirteenth and fourteenth centuries, but after that the rhythm swung back again. The average length of the cycle has been 100 years. The last center of the cold-dry phase occurred at 1865. One hundred years brings us to 1965. No one can seriously doubt that we must obviously be heading into what the forecast curve predicts. The center of the next cold-dry phase may jump ahead or back a decade, but it is not expected to do so this time. If it were going to do so, the signs would already be evident, but the signs are lacking.

"To be sure, the 100-year cycle, about which the Weather Bureau evidently knows little or nothing, does not repeat mechanically like a sound wave. It varies considerably in length from time to time, just as the sunspot cycle does. The sunspot cycle is not a fixed period; it varies from 8 to 16 years in length, yet no one hesitates to call it a cycle. The 100-year cycle is exactly as much a cycle as the sunspot cycle, including the cycle between major warm-dry periods. The 7-year rainfall cycle is as much a cycle as the sunspot cycle, as are the 6-year temperature cycle, the 18-year rainfall cycle, and the 60-year temperature cycle, and many others.

"Taken one by one, both the medium and the long cycles to be found in the great majority of the rainfall and temperature curves for the different states of the Union, and in the vast majority of the many individual stations used for this report, all point to the same forecast—lowering temperatures and declining rainfall. Also, masses of curves from weather stations of Great Britain, Europe, Asia, Africa, and Latin American countries show the same thing. When projected forward, they all drop sharply after 1950 to the cold and dry side of the median."

From this it can be concluded that prosperity and depression eras are predictable if the weather cycles continue to behave as they have in the past. Difficult times have occurred in each of the twenty-six preceding cold-dry phases of the 100-year cycle. We are now entering the twenty-seventh cold-dry phase since 575 B.C., and there seems no evidence to support hope for a time of economic prosperity. Although we cannot expect any different treatment from the recurrent cycle, we do have some knowledge of how to prepare ourselves.

It is Dr. Wheeler's contention that during the next twenty-five years, one third to one half of the years will be severely cold and dry. The entire period should average on the cold and dry side.

Already the growing season is shortening. In 1974 an early September frost destroyed millions of dollars' worth of maturing corn in the Midwest. Once the cold phase has stabilized, frosts in June and in August may occur in the corn belt. Early and late frosts will become a serious menace in southern California, Texas, Florida, and other parts of the Deep South. Winters will become so severe as to cause trouble for cattle and sheep raisers on the western ranges. Blizzards will become much more common than they have been for fifty years or more. Severe and lengthy droughts and famines will be worldwide.

The United States, and the world as well, should be preparing for long shortages in its water supplies and for shorter, not longer, growing seasons. We must prepare for colder weather and long droughts, which, in turn, will bring about scarcity of food for the prosperous nations, famine and starvation for the less economically stable. It would be wise, therefore, for agriculturalists to work on cold-resistant plants and faster-growing crops—especially faster-growing corn and wheat. Agricultural research should be concentrating on the production of larger numbers of cold-resistant, fast-growing crops if normal yields are to be maintained.

The northwest plains and mountain states will probably suffer the most, but the north-central and northeastern states will receive their share of bitter winters and shorter growing seasons. The most prosperous part of the United States, agriculturally, might become the southeast and Gulf states.

Frosts will penetrate deeper into the ground over the upper two thirds of the country, especially in the northern third. It would be well to make certain that water pipes are deep enough underground. There will be much more zero weather in the northern half of the country than during the last quarter of a century or more, and there will be more freezing weather in the southern half. John Cejka warns us that the winter of 1975-1976 may prove to be the most severe in recorded history!

Along with the fall in temperature will occur a serious decline in rainfall. This has happened consistently on the 100-year cycle, twenty-six times since the days of ancient Greece. Each one of these longtime drops in temperature and rainfall shows in the sequoia tree-ring curves. The longest of these curves goes back to 1350 B.C.

In the fifth century, near the time when Rome fell, world rainfall was so low that the California sequoias grew very slowly for many decades. The Caspian Sea in Asia sank forty-five feet below its present level. There are long cycles in rainfall just as there are long cycles in temperature. Just as in the case of the shorter cycles in temperature and rainfall, the longer ones follow, or parallel, one another. In the long run, temperature and rainfall fluctuate together. World rainfall is declining now, along with world temperatures. At the end of the century both will recover together.

Until then, however, agriculture and industry must prepare for serious trouble in regard to the national water supply. We are using water faster than nature can replace it. The water table is sinking alarmingly fast, everywhere. It will sink a lot faster and a lot farther in the decades to come.

It behooves the nation to store and to conserve every available gallon of water. We may be drawing on the Great Lakes and on the oceans for water before 1980. Irrigation projects of undreamed-of scope may become necessary. Remember that not only the United States and Canada will be hit, but the other continents as well. It will be necessary to develop drought-resistant, as well as cold-resistant, plants and crops. Temperatures may fall for ten years or more, then start climbing again, only to fall once or twice more before the next longtime climb begins, which is not expected to begin until around 2000.

Obviously droughts and frost will have their effect on our economy—but remember that it, too, is directly linked to the weather cycle.

The decline of an old world, such as we are experiencing in the next few decades, is inevitably characterized by a succession of rapidly occurring and troublesome depressions. Prosperity, as we have known it, is due to decline for an extended period. Times may have changed from the earlier terminations of 500-year cycles, but the laws of nature have not.

If cold-dry conditions begin our next twenty-five year period, there will be a temporary, but vigorous, period of prosperity around 1980. Economic conditions will reflect weather conditions—up and down.

There will be brief periods of depression, fleeting peaks of prosperity, many years of just plain leveling off, holding-one's-own kind of business survival methods. As rainfall increases, so, of course, will prosperity. However, the next era of prosperity is not expected to develop until around 1995. Severe and lengthy financial depressions will be worldwide. We must in the best ways we know how condition ourselves for recurring economic hard times.

The coming cold-dry depressions will not be all loss. According to Raymond Wheeler, they will be periods of the revival and the reestablishment of the free competitive system of economy that has always been destroyed by statisms and dictatorships and severe depressions of the preceding warm-dry periods. Primitive and backward countries will experience rapid advances, along with unrest, and will undergo accelerated industrialization. A major renaissance on a worldwide basis will occur as the next major warm-wet phase of the 100-year cycle gets under way around the year 2000.

18

Business, Labor, and the New Golden Age

We must expect the death of our old world to last until the end of the century. We must bolster ourselves with the thought that civilization has passed through four great convulsions since ancient times comparable to the one we are now entering. Each time the world has emerged better than it was before—more stable, richer, and with greater concern for individual rights. The period before us, although severely troubled economically, will witness great democratic reforms throughout the world.

"All this confusion is the reverse side of social reconstruction," Wheeler wrote. "Renewed faith in democracy takes root during these times. Emancipation of the underprivileged and oppressed occurs. The forgotten man is recalled and comes into his own as a *free individual,* not as a ward of the state! More important yet is a revival of religion and the spiritual power of whole peoples and nations, along with a shift from classical and abstract forms of artistic expression and literature to the romantic and concrete, extolling the importance of freedom and the individual."

We must accept a revolution that will expand and stabilize democratic institutions. This expansion will include the sharing by labor in the management and control of the economic and political processes of the civilized world. Now as we near the beginning of the sixth 500-year cycle, can we visualize what the democratic aspect of the coming revolution will bring to mankind?

Raymond Wheeler foresaw the rise of the laboring class to political and economic power.

"Heretofore," Wheeler tells us, "the working classes have had little to do with government and nothing to do with the means of production. The modern industrialized world is a product of the middle class. The American and French revolutions were revolutions of the middle class, not of the factory worker, peasant, or farmer. The modern world, thus far, has been a middle-class businessman's world."

Wheeler cautions conservative businessmen from becoming alarmed at what is really "a sign of progress." The demands being made by labor in our closing decades of the fifth 500-year cycle are only an extension of the principles of democracy into areas in which they have not previously been applied. "The new world will be the people's world." Where labor is not represented in the policies of industry, business methods will be liberalized and humanized.

Wheeler also acknowledges the potential danger of a dictatorship following a revolution of the proletariat, because the laboring class lacks the means, knowledge, and experience to construct a real democracy. "Freed from the tyranny of a monarchial form of government, chaos would reign without some new form of despotism to take the place of the old," Wheeler warns.

The high standard of living in the United States has produced better-educated working classes than elsewhere in the world, Wheeler observes. Therefore, "it is inevitable under these circumstances that they should want to share in running the economy of this country. They are rapidly acquiring the knowledge and the capacity to take a direct part in this process."

Wheeler firmly points out to management that there is no turning back the hands of the great cosmic clock. "The most constructive thing that management can do," he advises, "is to accept the inevitable. It must see to it that labor in this country does not fail to distinguish between false and real democracy—and that it does not ruin our entire economic system by making excessive demands due to lack of understanding and experience."

Within the next few decades, during the cold-dry period that has begun—and is due to last, with the usual interruptions, at least for the rest of this century—we can expect that labor will continue to apply pressure for wage increases, decreased per-week work hours, and increased representation in the councils of management. In the most advanced countries, such as the United States, Britain, and the nations of Western Europe, the working classes will keep striking until they share in the ownership and management of the business world. There will be some anarchism until the laboring classes become part of an extended "middle class." Business will become a cooperative enterprise between management and labor, until, perhaps, the familiar distinctions will disappear.

Judging by past activity in the revolutionary cold periods, the democratizing of business is inevitable. "The problem," we are instructed by Wheeler, "is to expand democracy by voluntary means, preserving democratic institutions and laws while the expansion process is being achieved. Now, during the next few decades, this new and powerful class of voters, the laborers, must, in a sense, be absorbed into the middle class and be given middle-class concepts of free enterprise and democracy. While assuming greater responsibility, they must experience success in helping democracy work; or else, when it turns warm again and centralizing and socializing trends get under way once more, there will be nothing but stagnation and ruin ahead. If this happens, the next warm period will produce a despotism as catastrophic for modern civilizations as were the final despotisms of Greece and Rome.

"By the proper emphasis upon education in our schools and by the proper cooperation between capital and management on the one hand and labor on the other, such a catastrophe can be prevented. During the next few decades when both the middle class and labor are democratically minded is the time to stabilize our institutions—enriched by the contributions and cooperation of labor—to the end that they will not collapse in the warm periods to come."

Down through history, the people who had control of the government and the economic system have resented and feared the necessity of sharing their power with additional classes of society. Although the social revolution of the working class may appear threatening to the middle class from today's perspective, Wheeler assures us that "the conflict between management and labor contains no necessary threat to society and will not culminate in socialism or communism. When viewed in the light of historical ecology, it is only the next and expected step in the evolution of true democracy. The net result of the revolution will, in the end, be greater opportunity and freedom for all classes."

"You see," John Cejka explains, "civilizations, like people, seem to get sick every so often.

"Some civilizations and cultures have become so sick that they have died or disappeared. By the time that happens, there is usually only a fraction of the original population left. That remnant wanders off and mingles with other societies—or else they are powerless to prevent other cultures from moving into their region and assuming control. In some cases, the death of a civilization leaves the whole area so devastated and disintegrated that the people left there revert to primitive, semisavage levels and remain there for quite some time. Eventually, though, a new society rises out of the ruins of the old."

Those of us who are living today are entering a cold-dry period and witnessing a time of revolution. Foresight and understanding will be needed by the civilized world if it is to survive the necessary changes without succumbing to catastrophe. No longer will nations be overrun by migrating hordes; however, as a result of wars in this century, the world has witnessed mass movements of peoples that far surpass in numbers any of the migrations of history. But out of the reshuffling and societal confusion will come a new epoch in history, beginning by 2000, bringing with it a renaissance, unprecedented prosperity, and world government. This collective rebirth will involve the great masses of the people participating in the economic and political structures to an extent unknown before in history. As a consequence of the revolution, Wheeler stated, the prevailing political structure will be neither fascism nor communism, but a purer, truer, more effective and complete democracy than has ever been known. Those of us who will be alive at the termination of this coming time of famine, revolution, depression, and the collapse of the old civilization, will witness the advent of a new Renaissance, a surge of renewed life force, a restructured political and economic viability.

The turning point will come some time in the 1980s. Right now, according to the research of Raymond Wheeler, we are living at a time comparable to 1480, just before the advent of the great Renaissance of 1500.

"What if the cycles should suddenly be altered and the next thirty or forty years should prove to be warm, rather than cold?" I asked John Cejka.

"In my opinion," Cejka said, "even though the next twenty-five years are going to be very rough, a period of warmth would be the worst possible thing with our civilization as sick as it is right now. Assessing Raymond Wheeler's detailed study of the rise and fall of civilizations to longtime weather trends, we would probably end up the way the Greek city-states did when they attempted state socialism—or we would probably sink into almost total destruction the way Rome did after she tried to dole out sustenance to the entire known world.

"You see, socialistic trends develop during warm cycles, especially the hot drought periods in the 100-year cycle. The revival of individual responsibility and an emphasis on free competition, private enterprise, and personal freedoms always occur during cold times. Democracy revives during cold times because man has more energy then. Not only is his drive and ambition greater, he is more aggressive, more optimistic, more willing to take risks and work hard. Cold times are revolutionary times. New ideas and methods can be more freely introduced in the cold periods.

"But it is not going to turn warm. It has definitely turned cold already, and it should remain so until at least the end of this century. But I am not saying that a revival of democracy and individual responsibility is assured.

"First of all, democracy must be adjusted to the new age which emerges out of the ruin of the old civilization.

"Secondly, there must be men and women with the energy and determination to create a democratic revival when nature has supplied them with the vitality and inclination to do so. If too many people slump into an attitude of pessimism and indifference toward our societal welfare and maintain a sense of despair while the old world is dying, there will probably be enough of a socialistic foundation left over from the previous warm period to permit totalitarianism to engulf us before anyone realizes what is happening . . . and before the next warm period has scarcely begun."

That all this should happen is unthinkable. "The human race, dull and stupid as it is, greedy and selfish as it is, is not that stupid or that greedy or that selfish. It will see the handwriting on the wall," Dr. Raymond Wheeler declared. "Man will stabilize his institutions; fight the war to end war (which may be the next one); work out a democratic, internationally tolerant and cooperative society of nations, which will bend their energies toward constructive pursuits; discover at last that there is no progress in destruction. He will not do this on moral grounds. *Nature will compel him to do it—or he will not survive.*"

"At the end of one of these 500-year rhythms," John Cejka said, "civilization *needs* a thorough reorganization. In the past, each time one of these cycles evolved, civilization emerged from the chaos and confusion in much better condition than it was before the terrible sickness began. I am confident that we can do it again."

"Is our recovery in the cycles as well?"

"Yes, but the recovery must be man-made, man-created, and

man-directed. In the rough days ahead of us, we need leadership, patience, and endurance so that we will be able to restructure our civilization along new lines."

We must keep calm, preparing ourselves for more, rather than less, confusion.

19

Free Will, the Stock Market, and the Cell Barometer

By now, perhaps, the reader is becoming somewhat uneasy. Can it really be that all of man's magnificent accomplishments, all the fine products of his reason—together with his animalistic, insane behavior which results in wars and riots—are actually due to energy waves, cyclic forces beyond man's control?

Can it be that one level of man's consciousness is attuned to something like electromagnetic impulses, which even though he cannot hear them or define them, do dramatically affect him and cause many of the crises of existence to which he is subjected?

No thanks, the reader may say. Such a theory is too deterministic. I am master of my destiny; I have freedom of choice.

In the book which he wrote with Og Mandino (*Cycles: The Mysterious Forces That Trigger Events*), Edward R. Dewey states that a "basic secret of nature" is that man is surrounded by cyclic forces which "bounce us like marionettes on a string." Dewey admits this concept is unsettling.

"Since it is demeaning to his self-esteem," Dewey writes, "it is perfectly understandable that man should resist any hypothesis that holds that his life and his universe vibrate in rhythms that are regular and at least partially predictable and are caused by a force or forces still unknown and possibly uncontrollable. . . .

"Nevertheless, the evidence that man is not one step down from the angels, sublimely in command of himself and his world, continues to accumulate. He is more like a character in a Punch and Judy show, pulled this way and that by environmental forces. And he will continue to be so manipulated until he solves the mystery of these forces. Only then will he be able to cut the strings and become himself."

"Again, Brad," John Cejka reminded me, "I keep likening our charts to a roadmap of time. Our charts deal with the fixed distances in space between all points in a state. The time it takes you to get from one point to another depends upon how fast you can travel, right? You need to know the speed or the rate of motion and the time—these are your variables. The distance you travel is fixed.

"Our weather maps are actually time maps. Time is a fixed quantity. The variables are space and motion. Understanding this, we can predict the future effects of weather, just as you can predict the time it takes you to drive from Decorah to Chicago."

The Wheeler Culture Curve and the Maxwell Weather-Energy Cycle enable man to snip himself free from puppet status. When one possesses a Roadmap of Time, he may not be able to alter curves in the highway, but he can prepare himself by knowing several miles away when the sharp bends will appear.

Perhaps man cannot control nature, as he has for so long smugly believed, but he may certainly adapt himself to nature's whims and its cyclic forces. The American Indian learned this basic truth of material plane existence generations ago, and he has never been able to understand the European's arrogance toward the Earth Mother.

Cyclic forces are inevitable, but by knowing about them in advance through the Wheeler-Maxwell discoveries, we may exercise our freedom of choice and hold the cyclic result subject to our will. Knowledge of what is ahead for us on the recurring pattern of time can enable us to prepare for adversity.

To present an admittedly simplistic example, say a group of men and women are suddenly transported from an area of perpetual warmth and sunshine to a land of seasonal change. Since they arrive in the summer months, they soon make their adjustments and begin to settle in what is to them familiar patterns of existence. But they do not know about winter. Their casual shelters offer no protection against freezing winds and snow. Their clothing is inadequate. Their meager food stores are soon exhausted. Quite likely, they would all starve and freeze to death.

If we choose to structure our lives according to a linear unraveling of time, we will be as unprepared for the onslaught of crisis situations as were the unfortunate immigrants of our above example. If we can recognize the cyclic forces and their rhythmic intervals, we will be in the position of those who know that winter is coming.

There is nothing we can do about the cycle that creates winter, but we can prepare for the icy months by making warm clothing, winterizing our homes, and putting up great stores of foodstuffs.

Cejka got up from his chair, not to pace in frustration, but to reach for a thick volume of Wheeler's notes which were heavily laden with Maxwell and Wheeler's research data. He flipped through the bound pages of the book until he found the right page.

"Listen to this," he began. "Here we have an answer to your 'free will' problem in Wheeler's own words."

> [Man] fought the idea that the earth was not the center of the universe, not so much for religious reasons . . . but because he resisted the idea that his abode, especially to himself, was not the center around which the universe revolved. Steadfastly, in spite of a vast literature to the contrary, man has refused to place himself on the same plane as plants and animals in the great science of nature. . . . He will not take kindly to the fact that his behavior is extraordinarily sensitive to climatic influences. It will seem too

> much like surrendering intelligence and will to inexorable and mechanical laws.
>
> But this, again, will be a sheer rationalization. Man's dependence upon the laws of gravitation, of electricity, of the gas laws, of the laws of health and disease, has been the source of his emancipation—not the source of a fatalistic environmental determinism. Man's dependence upon climate and climatic changes will become an added source of freedom if and when he will submit himself to the facts and turn those facts to his advantage. So long as he remains ignorant of these facts, he will continue to be a victim of the forces to which they point; but the instant he accepts the facts and proceeds to do something about them, his status as a victim will be transformed to the status of an intelligent agent—cooperating with the forces of his environment, rather than allowing them to govern him.

"That sounds very much like the philosophy of the American Indian, that one should learn to live with the Earth Mother rather than try to control her or fear her," I commented. "It also seems Wheeler was very much tuned in to the concept of the whole; that all living things are one with another."

Cejka nodded, smiling the way a teacher does when one of his slow students is beginning to catch on.

"Maxwell found that there are high death periods tied to the weather. He found that more men die when weather conditions are different from those prevailing when more women die.

"Elderly women pass away when there are intense variations or changes in air pressure. Men tend to die when there has existed a long period of low pressure. Maxwell placed the reason for this on the 'cell barometer.' He held that every part of the human body is made up of cells which are self-contained tiny universes," Cejka continued. "Their outer skin may change shape, but, normally, remain the same size. He theorized that since low pressure must force someting out of the cell for it to retain its size, it must have an expulsion of gases. Cells are made up of liquids, gases, and other matter. Low barometric pressure causes an expansion, and since the body does not swell up during low-pressure periods, something has to be expelled from the cell. That something is gas, and part of that gas is oxygen. The cell, and therefore the human body, expels oxygen and other gases during low-pressure weather conditions.

"So Dad goes out to shovel snow, and he compounds a negative health condition. If he should get that heart attack, he is rushed to the hospital and given *oxygen*. But because the body is busy expelling oxygen during this low-pressure period, it simply cannot accept the forced oxygen. That poor man's body is going to have to work even harder under stress conditions to expel the oxygen they're pumping into his lungs. The only safe way to establish the oxygen balance would be to administer oxygen in an air-locked room, so as to isolate the treatment room from outside air pressure. Maxwell always maintained that operating rooms, recovery rooms, and intensive-care units should be air-locked.

"I've verified the cell-barometer theory in my work with freeze-dried seeds and in other agricultural experiments. It could be that the cell-barometer theory may be one of the most vital of all Maxwell's legacies. Through Maxwell's discovery, each man and woman can chart his or her own personal cycle with astonishing precision. Months in advance, you can map out which will be your good days and your bad days. You can know when to plan a trip, when to enter surgery, when to have a baby."

"Could you explain just a bit further about each of us being able to chart his own personal cycle?" I asked.

"It can be as elaborate as you want," John replied. "Just try a simple chart at first.

"Take a few minutes each evening to review the day just past. Was it a good day or a bad day in terms of your general mood, attitude, emotions? Were you depressed, happy, excited, sad, worried, grouchy, irritable?

"Draw a chart on graph paper and place a dot on the appropriate spot each day. Pretty soon, you know, you will have a regular, dot-connected chart which will indicate your ups and downs. After a month or so, you should begin to see a pattern emerging which will represent your natural mood-attitude-emotion rhythm. A few months more, as additional personal data accumulates, you should be able to peek into your own future and forecast your highs and lows. As you acquire greater knowledge concerning your personal cycle—and therefore yourself—you will see that even tough 'downer' days will pass. You can see where you can change certain things—such as not scheduling a party, a dental appointment, or an important meeting when you will be in a low period—and prepare to face things which cannot be changed, such as that social engagement or business appointment scheduled by someone else."

Another question occurred to me. "All right. If we can actually have accurate foreknowledge of future happenings on the time cycle—and that includes the stock market—I am going to set about amassing a large personal fortune. But if I can do it, what is to prevent every reader of this book from using the principles outlined here and doing the same thing? Surely such foreknowledge would seriously disrupt the very cycles which one is attempting to exploit, and the market would certainly collapse."

John Cejka answered my question in this manner:

"It is true that anyone possessing this specialized data will be able to predict the curves of the stock market. But even though anyone acquiring this information has the *potential* to become wealthy, his own psychological and his own individual cell-barometer curves, among other things, will determine whether or not he will be able to use the information correctly.

"Many of our clients have asked that same question. 'If all this information gets out and everyone starts following the cycle, will the cycles still be there?'

"Because all human events enter into these cycles, all that would happen is that the maximum altitude of the cycle would be deemphasized. We wouldn't lose a cycle at all, we would merely smooth it out. You cannot defy the law by which you made the prediction.

"Let us take the commodities market as a specific example. Remember one fundamental: The cycles are made by people who have participated in the market. It is a law of the marketplace that 90 percent or better of the traders lose money. You are not going to defy the law of time itself. You are using the past to predict the future. If you make one hundred forecasts for one hundred individuals, I will guarantee that 90 percent or better will still lose money.

"The marketplace is always two people—the buyer and the seller. The professional trader in Chicago always takes the opposite side of what the public wants to do. The public is always wrong. When the public becomes convinced that the market that is 90 percent bullish is going to go up, the next day it will collapse. That is mass mind in function.

"When I first began market trading, I heard that the market becomes so erratic that it becomes an absolute dilemma. But I saw that was because every trader pits himself against the market. He is in there not necessarily to make money. It is King of the Mountain stuff. Your intelligence above everyone else's. You can get eaten up in there just as if you were in a jungle.

"But then I realized that the male mind cannot understand the marketplace. It is as fickle as a woman. The market is always a reflection of two—buyers and sellers, positive and negative. The male, predominantly, is the one who plays the market. The market is always making an opposite reflection. The market is the Big Woman, and on some dimension of reality exists as a being. The professional traders are its babies, and the Big Mamma feeds them with the sustenance that the losers surrender.

"That may be figurative speculation on my part, but the point I want to make is that other factors—including very definite psychological factors—enter into these cycles. Human nature has not changed in all these thousands of years. That is why I count on these cycles.

"If mass mind were to will away such things as droughts and famines, they would probably cease to exist. But until mass consciousness has been elevated to that degree, such things as droughts will continue to occur. What we can do now is to utilize the weather-energy cycle to *predict* such occurrences.

"If you study these cycles and if you really believe that they truly predict the climate conditions of the future, then you will know that, for example, a drought really will occur in such-and-such a year. Once you know that a drought will occur, you can always take steps to protect yourself. As Abraham Lincoln once said, 'If you know where you have been and if you know where you are, you surely should know where you are going.' Maxwell and Wheeler found out where we are going by correlating all human affairs with the weather."

Edward R. Dewey writes: "The study of cycles *can never give complete foreknowledge*. There will always be accidental variations and noncyclic factors that will enter into every situation, no matter how much we know about cycles."

Dewey points out in regard to the stock market that nearly every transaction involves one correct and one incorrect forecast of the future price of that item. Nearly every commodity price, stock price, or whatever else is going to go either up or down in the next hour, or next day, or next week.

"Whenever a trade is made, one of the two parties involved, either the buyer or the seller, has guessed wrong . . . despite all the facts and advice . . . available," Dewey observes. "We will improve our results as we learn more about our mystery and its cause."

So while it is theoretically possible for *everyone* to interpret the cyclic forces which control stock market fluctuations, not everyone will—due either to misinterpretation of the data or to the influence of his individual cycles.

20

On Becoming Integrated With the Cyclic Environment

Around 2040, perhaps a bit earlier, a heat climax will occur that will exceed the recent one of the 1930s. This might well be the apex of the current 1,000-year cycle. But after the heat climax, glaciers will begin to advance again, and the world's temperatures will fall until a cold climax is reached around 2500. At that time, the world will experience another upheaval as complete as that which terminated the Middle Ages and inaugurated our modern world.

When a scientist is dealing with facts and principles that tap human sentiment and emotion, and these facts and principles, if true, require an adjustment in human philosophy, he should be willing to share in the making of that adjustment.

"I have always insisted that the outlook for man is not fatalistic, integrated though he is with his environment and subject profoundly to climatic influences," Dr. Wheeler stated. "There is no excuse whatsoever for becoming an environmental determinist—at least of the kind known to the history of science."

Naturally, everything that happens in this world must have a reason. The science and philosophy of the past have given to humanity differing points of view regarding reasons for events. One world view supposes that the universe operates in accordance with blind, mechanical laws, that the forces of nature are physical (in the sense of material forces), based on matter as the ultimate reality.

"If this philosophy is carried out consistently to its natural conclusion, as certain philosphers and scientists of the past have done, there is no place for the mind in such a world; there is no place for human values; there is no place for God," Dr. Wheeler said. "The universe is a vast, meaningless machine, having no purpose, going nowhere. Reality in the end is but a fortuitous concourse of whirling atoms. Laws and principles mean nothing except for purposes of prediction. The human organism, like everything else in the world, is but a mere machine, a robot, whose feelings, thoughts, and longings have no more significance than the bumping of one atom into another. Human life has no more purpose, no more value, than a stone, among millions of other stones, clinging to the side of a vast mountain.

"The fact that most scientists will not carry such a view through to its logical conclusion, and that there have been but few philosophers down through history who were willing to do so —although they entertained a materialistic and mechanistic view of nature for everything outside of human affairs—is sufficient to prove the emptiness of the view itself."

Although the critical mind has not always found it easy to offer a substitute, scientists and philosophers who have provided an alternate view have outnumbered the first group several to one all through history. The second means of apprehending one's environment is the organic and idealistic view, as opposed to the mechanical and materialistic one.

The organic view maintains that the laws of nature are not

wholly mechanistic. Predictability and regularity do not imply a fortuitous operation of mechanical forces. Neither man nor the universe is a machine. The term "material" has no meaning when applied to physical forces; the term "matter" has no meaning when applied to the world in which man lives—or when applied to man himself. In fact, the term "physical" has no meaning other than a superficial and convenient one. It is merely a descriptive term which conveniently distinguishes certain types of phenomena—or certain types of events—from others just as the term "robin" points to a class of birds to be distinguished from "pheasants."

The organic view is a positive one. The universe means just what the term implies: *uni*, "one," and *versum*, "combined or turned into." The universe is One, a Unity. This means that all its parts, from the humble atom to man himself, are interrelated in a common whole; everything in the universe obeys the same ultimate laws.

But if there were not something to the universe in addition to its separate parts, these parts would not be interrelated; there would be no whole; there would be just the parts. In this sense the whole is oversummative or supersummative. It is more than the sum of its parts. Many different terms have been used to describe this "something more": organization, relations, integration, system, preestablished harmony, field property, plan, design, intelligence, world soul, God.

"Since the universe is a vast Unity, each part in it from the smallest to the largest, the simplest to the most complex is related to every other," Dr. Wheeler wrote. "This means that whatever it is, and whatever it does, each part has a function to carry out in the 'plan' of the whole.

"At the human level of existing, things that function may legitimately be called purposeful. Since the whole transcends the parts, whatever the ultimate nature of reality, there is room for God—in fact, many philosophers and scientists today would identify God with the oversummative character of the natural world. This makes God transcendent and superhuman, but not necessarily supernatural. It does not say what God is, except to point out that there is no reason for not identifying God with the plan of the universe that gives to each object in it a function, a purpose, which it carries out in the scheme of the whole.

"Modern science has revived the ancient concept of teleology (the view that everything in the world has a purpose), but in clothes so different that no one except the philosopher would be likely to recognize it. According to modern science, no activity can transpire in the world, there can be no motion of any kind, unless nature as a whole obeys laws of equilibrium. Activities have *direction*, or they cannot occur. They are headed somewhere. There is activity only when there is an assumed difference in potential between two points in a unified system of energy. Then the activity proceeds from the position of the higher to the position of the lower potential until the difference between the two is resolved; that is, until the system reaches equilibrium. Thus, while activity is going on, the state of equilibrium is in the *future*, but the events that are now occurring are doing so with reference to that future state. In other words, the conditioning factor—or at least one conditioning factor that is assumed to be necessary for the action—lies in the future of the events in question. The event is headed toward the goal of future equilibrium. The purpose of the event is the restoration of the system to equilibrium.

"One could go further and show that, by definition, this equilibrium is an ideal state. Absolute, permanent equilibrium in the universe would mean the end of all activity. The goal is always ahead; it is never reached, but gives direction, always, to present events. In human affairs that 'direction' gives a purpose to the individual life and to collective living.

"Thus it is that human behavior can follow the same ultimate laws as any other type of event in nature, without jeopardizing human values. Indeed, everything in the universe has to follow the same laws, or the universe would lack unity; there would be no integration, there would be no universe. Human ideals are objectives, utopias, perfections of one kind or another, that, by definition, lie ahead of human action. If these ideals could be reached, they would not be ideals. Like the condition of perfect equilibrium of all the physical forces in the universe, they are forever in the future; but that future is forever giving direction to what is happening now."

To become integrated with the environment does not mean that man will exist in a planless, mechanical world. It is convenient to define physical laws as laws of dynamics, as opposed to laws of mechanics (laws of dynamics involve the motion of forces resolved toward equilibrium, or acting and interacting under laws pointing eventually to principles of balance as the basic principles of nature). To say that human behavior follows laws of dynamics, the same laws which are obeyed by physical forces, is not to say that man is reduced to a machine (or to the level of the physical atom), because the view of integration denies that the atom obeys mechanical laws. In the organic view there are no such laws.

"It is neither elevating the atom nor degrading man to assume that every object in the universe obeys the same laws," Dr. Wheeler assures us, "for the status of each object is defined, not by the laws it obeys, but by its function, or purpose, in a system where every part obeys the same laws of the whole. The functions or purposes of the two may be vastly different under the same laws.

"The human being has intelligence, which the atom or the rain or the heat or the cold does not have. Intelligence has a place, a purpose, in the world. What better purpose could it have than that of trying to ascertain the meaning of life and of the universe as the abode of man? What better purpose could it have than that of searching for the truth?"

Modern philosophy and science alike have come to the conclusion that an effect is never the passive victim of a cause. Under the law of action and reaction, for example, there could be no effect of a cause unless, in turn, the object acted on was itself active in a direction counter to that of the cause.

Pounding one's fist against the table is a good example. The table must hit the fist just as much as the fist hits the table, or there would be no impact. Nothing in the world is passive; everything is active, even in cause and effect relations.

Human qualities, then, are not passive victims of environmental forces. Human behavior is a force acting *against* environmental pressures, just as much as environmental pressures are forces acting upon man.

Intelligence is one of these qualities, one of these modes of behavior. Intelligence, then, is an active agent in the world, capable of producing effects under natural laws—indeed under the same laws "obeyed" by forces acting on intelligence, many of which come from climate.

"When intelligent behavior is consciously and purposely guided by the results of experiment, discovery, and insight—even though its existence depends upon environmental forces—it can, by insight and knowledge, determine by its own choice the manner in which environment shall effect it," Dr. Wheeler said. "Knowing that in the past the excess energy that the human race has possessed under certain climatic conditions has led to destructive aggressiveness, man can determine that that aggressiveness be expressed in a constructive manner, which will result in happiness rather than in suffering; in life rather than in death. There is nothing about climate that says, 'Thou shalt kill.' It merely has determined the energy level (aggressiveness) that, in his stupidity and ignorance, man has used for purposes of killing."

There is the possibility that man will eventually be able to create enough of an artificial climate in which to live to keep him from going to extremes under the influence of natural climate. By air conditioning, for example, he may be able to reduce in considerable measure the lethargy on the one hand and the fanaticism on the other which have been associated with hot drought epochs. He may be able to minimize the heretofore uncontrolled belligerence that has been associated with climatic transitions. He may become enlightened enough to eliminate the civil wars of colder times.

Man can improve himself physically as well as mentally; he can maintain a more even health, a more even birthrate, a more even and continuous occurrence of good leaders and geniuses. He can eliminate periods of neurotic and fanatic leadership and followership, and he can maintain a more stable economic system.

By submitting to the gas laws man built engines which were

capable of carrying him around the world on land, sea, and air.

By submitting to the laws of electricity, man gave himself light and heat, and instruments of communication, which could bring the people of different continents together as if there were no time or space.

By submitting to the inevitable facts of health and disease, man has reduced infant mortality, prolonged his life, conquered devastating plagues, enabled the lame to walk, the deaf to hear, and the blind to see.

Why, then, should man not control his destiny in larger ways—in part by a regulation of the climate in which he lives—and rid the world of excessive aggressiveness and sloth, of overambition and decadence, of tyranny and anarchy, of slavery and war? Let man apply his scientific discoveries and insights into large-scale programs of human betterment, not alone in the individual home, the small community, but also on a national and international scale.

Wars will rage, and governments will continue to collapse in internal misery and bloodshed until the natural causes of these events are known and recognized—even as today man knows and recognizes why it is warm in summer and cold in winter.

"The end of human progress is not in sight," Dr. Wheeler pronounced with an optimism predicated on a careful assessment of his research data. "There is emancipation ahead through a control of human nature, even as there was emancipation ahead when it became possible to control forces in the physical and biological world.

"Is there any intelligent person who, by interests defective in their narrowness, would delay the hour when there shall be peace and security for all mankind?

"Is there any nation who, by customs that are absolute, by ambitions that are collared in their selfishness, and by a regard for human life so low that success will remain a synonym for brutal dominance, would delay the hour when there shall be peace and security for all mankind?

"Is there anyone who will not hope and work for the day when never again will the days of Tamerlane, Hitler, and Stalin be revived?"

Index